An Introduction to Star Formation

Guiding the reader through all the stages that lead to the formation of a star such as our Sun, this textbook aims to provide students with a complete overview of star formation. It examines the underlying physical processes that govern the evolution from a molecular cloud core to a main-sequence star, and focuses on the formation of solar-mass stars. Each chapter combines theory and observation, helping readers to connect with, and understand, the theory behind star formation. Beginning with an explanation of the interstellar medium and molecular clouds as sites of star formation, subsequent chapters address the building of typical stars and the formation of high-mass stars, concluding with a discussion of the by-products and consequences of star formation. This is a unique, self-contained text with sufficient background information for self-study, and is ideal for students and professional researchers alike.

DEREK WARD-THOMPSON is Deputy Head of the School of Physics and Astronomy at Cardiff University. An observer in the field of molecular clouds and protostars, Professor Ward-Thompson's research interests lie in observing the formation of stars and planets, particularly the very earliest stages of star formation.

ANTHONY WHITWORTH is a Professor in the School of Physics and Astronomy at Cardiff University. Professor Whitworth's main area of research lies in the theoretical modelling of the formation of stars and brown dwarfs.

This book is based on lectures given by the authors, at Cardiff University and elsewhere, on star formation.

An Introduction to Star Formation

Derek Ward-Thompson
School of Physics and Astronomy,
Cardiff University

Anthony P. Whitworth
School of Physics and Astronomy,
Cardiff University

CAMBRIDGE
UNIVERSITY PRESS

CAMBRIDGE
UNIVERSITY PRESS

University Printing House, Cambridge CB2 8BS, United Kingdom

One Liberty Plaza, 20th Floor, New York, NY 10006, USA

477 Williamstown Road, Port Melbourne, VIC 3207, Australia

314-321, 3rd Floor, Plot 3, Splendor Forum, Jasola District Centre, New Delhi - 110025, India

79 Anson Road, #06-04/06, Singapore 079906

Cambridge University Press is part of the University of Cambridge.

It furthers the University's mission by disseminating knowledge in the pursuit of education, learning and research at the highest international levels of excellence.

www.cambridge.org
Information on this title: www.cambridge.org/9780521630306

First published 2011

A catalogue record for this publication is available from the British Library

Library of Congress Cataloging in Publication data
Ward-Thompson, Derek, 1962–
An Introduction to Star Formation / Derek Ward-Thompson, Anthony P. Whitworth.
    p.   cm.
Includes bibliographical references and index.
ISBN 978-0-521-63030-6 (hardback)
1. Stars – Formation.   I. Whitworth, Anthony P.   II. Title.
QB806.W36   2011
523.8′8 – dc22   2010042727

ISBN 978-0-521-63030-6 Hardback
ISBN 978-1-107-48352-1 Paperback

For Jane, Hilary and the boys

Contents

Illustrations

Preface

This book is directed at the student undertaking a course in star formation for the first time. This may be in the later years of an undergraduate degree in physics, astrophysics, or physics with astronomy. Alternatively, it may be that the student only meets this subject for the first time during the first years of a masters degree. In either case we have assumed that the student already has a grounding in physics and mathematics, including, for example, Maxwell's equations, quantum mechanics and the laws of thermodynamics. Nevertheless, we find from teaching experience that brief reminders to students of things they learnt in other courses are generally welcomed as helpful. Hence, we remind the reader of some of the important points from other branches of physics where they are relevant.

We assume only a minimal knowledge of astronomy, and we derive the necessary astrophysical equations as we go along. We assume no prior knowledge of the subject of star formation itself and begin from first principles. Throughout the book we attempt to stay on ground that is firmly established, and try to avoid that which is trendy or the latest discovery. Experience has taught us that these matters often become outdated much more quickly than the solid foundations on which the subject is based. In cases where we stray onto less sure footing, we inform the reader that we are doing so.

The book does not aim to be a comprehensive encyclopedia of star formation, but merely an introductory text, as the title suggests. The biggest problem when compiling such a work is knowing what to leave out. We have tried largely to include topics that lend themselves to mathematical demonstration, even if that leads to slight over-simplification of cases encountered in the real Universe in this very complex field. We therefore apologise in advance if we have omitted any reader's favourite topic or detail. However, we hope that the reader will nevertheless find the book useful.

The ordering of the book is that we first assemble the necessary tools, and then we cover all aspects of star formation in the order in which they occur for solar-type stars in an evolutionary sense. Then we look at some of the ways in which higher-mass stars differ from this picture. Chapter 1 sets the scene with some introductory and background material. Chapter 2 discusses the electromagnetic radiation that we receive from star-forming regions, and how we use this to discover the physical properties of those regions. Chapter 3 looks at the interstellar medium, where the raw materials exist for the formation of future generations of stars. Chapter 4 studies molecular clouds, where the majority of star formation takes place, to discover the initial conditions for star formation.

Chapter 5 describes the issues associated with collapse and fragmentation on the way to forming a star. Chapter 6 covers the growth of a star from the seed of a protostar to a main-sequence star of roughly solar mass, through its pre-main-sequence evolution. Chapter 7 examines some of the issues peculiar to higher-mass stars and the effects they have on their surroundings. Finally, Chapter 8 gives a few 'tasters' of subjects that flow from star formation, which will hopefully lead the reader into further related topics.

There is an index as well as a list of symbols, to aid the reader. Where possible we have tried to avoid the use of the same symbol for two different meanings. However, we have also tried to use the symbols that are most commonly used in the scientific literature, so that the student is not lost when moving on from this book. Occasionally this leads to clashes. So we have made it clear in each case, when defining every symbol, what meaning we are using for that symbol, and wherever possible we have used a different font or subscript to remove ambiguities.

Our aim is that a student who has read and understood this book should be ready to undertake a higher degree in this field, to read and understand more advanced research texts in the subject, and to embark upon research of their own.

There are many people we would like to thank, who helped in the fashioning of this book, including many students, both undergraduate and postgraduate, who have given helpful feedback and comments on the text. We wish to thank a number of our postdocs, who have also read the text and commented on it, including Annabel Cartwright, Jason Kirk, David Nutter and Dimitris Stamatellos. We would also like to thank Peter Brand, Shantanu Basu and Jonathan Rawlings, who each read and commented on parts of the book, although any mistakes that may remain are entirely our own. We wish to thank Cambridge University Press for their patience, especially Simon Mitton, Adam Black, Jacqueline Garget, Vince Higgs and Claire Poole. Finally, we wish to thank our wives and families for putting up with us!

Derek Ward-Thompson
Anthony Whitworth
Cardiff
March 2010

Chapter 1
Introduction

1.1 About this book

It can be argued that astronomy is the oldest science. Since pre-historic times humans have gazed at the night sky and wondered about the nature and origin of stars. We now believe we understand a great deal about the nature of stars, but many aspects of the origin of stars remain the subject of intense study to this day.

In this book we aim to introduce the reader to the fundamentals of the subject of star formation. We describe the background physics underlying theories of star formation, and take the reader to the frontiers of current knowledge of this subject. However, we will make clear as we go along the points where we reach material that is less well established.

One of the most fundamental observations in astronomy is the fact that the night sky appears to be full of stars. Yet the processes which lead to the formation of those stars have taken astronomers many years to work out. Unlocking the mysteries of star formation has required the use of new techniques and the opening of new wavelength regimes to astronomy. We describe the chief physical processes which are believed to be important for star formation, and point out the role which each branch of observational astronomy has played in solving the various problems associated with star formation.

In this chapter we begin by introducing some of the main constituents of a galaxy, namely the stars, the medium between the stars and the gravitational and magnetic fields. We discuss their spatial distribution, and introduce the life-cycle of a star and the way in which the formation

Fig. 1.1. A rough sketch of the Hertzsprung–Russell diagram, illustrating the main sequence, where a solar-type star spends the majority of its life.

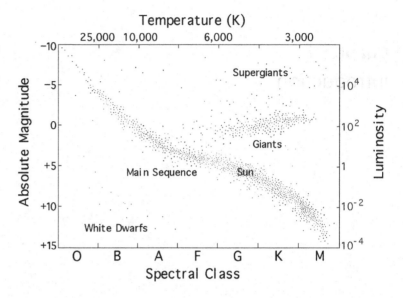

of a star fits into this cycle. We introduce the sites where stars are formed and give a description of the initial mass function of stars, the explanation of which is one of the major challenges for any star-formation theory. We finish with a list of some of the chief objectives of star-formation theory.

1.2 The stellar life-cycle

The most important diagram for stellar evolution is known as the Hertzsprung–Russell (HR) diagram. The HR diagram plots the luminosity of a star against its colour. In this diagram it is seen that the majority of stars lie in a single strip along the diagram, known as the main sequence, wherein the brightest stars are also the bluest, while the faintest stars are the reddest.

Figure 1.1 sketches an HR diagram to illustrate the approximate position of the main sequence, and the position of the Sun, which is simply an ordinary main-sequence star. In the Universe today, a star of roughly solar mass spends the majority of its life-time as a main-sequence star, during which time its energy source is the fusion of hydrogen to helium, deep in the core of the star. A star that has not yet reached the main sequence on the HR diagram is known as a pre-main-sequence star. However, before a star reaches the main sequence it has to be formed from the material in interstellar space. This involves a series of stages of contraction and growth by accretion under the influence of gravity.

Later in its life, when the majority of the hydrogen in its core has been processed by fusion into helium, a star leaves the main sequence, expands to become a giant or supergiant star, and undergoes various stages of losing mass. The most violent and best known of these mass-losing stages occurs for the most massive stars, and is known as a supernova explosion. However, lower-mass stars also undergo phases when they eject material. This ejected material can then form some of the ingredients for subsequent generations of stars.

Hence we see that stars undergo a life-cycle in which new generations of stars are formed in part from the debris of previous generations of stars. In fact almost all of the material which forms a new generation of stars, apart from hydrogen and helium, is the product of fusion in the centres of stars of previous generations. Thus the formation of stars is not the beginning of a linear process, but is an integral part of a cyclic process. The subject of this book is that part of the stellar life-cycle which occurs prior to the main sequence. In this book we start at the birth-place of stars, and follow the progress from there all the way to the point at which a star joins the main sequence.

1.3 The space between the stars

The space between the stars is known as the interstellar medium, or the ISM. The principal constituents of the interstellar medium are matter (gas, dust and cosmic rays), electromagnetic radiation, a gravitational field, and a magnetic field. The composition of the matter is typically 70% hydrogen, 28% helium, and 2% heavier elements, such as oxygen, carbon and nitrogen. Most of the matter – around 99% – is in the gas phase.

The gaseous ISM can be modelled with four phases: regions of very diffuse, hot, ionised gas which form a network of interconnecting stellar-wind bubbles and supernova remnants; a warm, partially ionised gas filling most of the rest of the volume of the Galactic disc; small clouds of cool, neutral, mainly atomic gas; and larger clouds of cold, dense, mainly molecular gas.

These larger clouds are known variously as 'dense clouds', 'dark clouds' or 'molecular clouds', depending on the context. Even in the centres of these clouds, the densities are only about 10^{12} molecules per cubic metre, so the term 'dense cloud' may be somewhat misleading. However, this density is still much greater than the average ISM, where there are typically only about 10^6 atoms per cubic metre. The largest clouds of molecular gas are known as 'giant molecular clouds' (GMCs),

Fig. 1.2. An optical image of the dark cloud Barnard 68. Note how the background stars are not visible through the cloud.

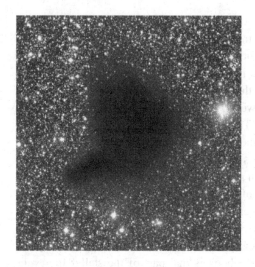

or GMC complexes, and can contain masses up to a few million solar masses[†] and extend for tens of parsecs.[‡]

The term dark cloud arose because at optical wavelengths these clouds can be completely opaque. Figure 1.2 shows a picture of a dark cloud taken at visible wavelengths. Note how the light from the background stars, which can be seen across this field, is obscured by the material of the cloud in the centre of the image. However, modern astronomy utilises many more wavelengths than the optical. The infrared and radio wavelength regimes have become more important than the optical in studying star formation, as the regions in which stars form are not so opaque at these longer wavelengths.

The reason that clouds are dark is that they contain not only gas, but also dust, which is opaque to visible light. This dust consists of very small grains, i.e. particles of solid matter, less than a micron in size, and consisting mostly of silicates (sand), and carbon compounds, probably including graphite. Dust grains have similar sizes to the particles of cigarette smoke. By mass, the dust grains only represent about 1% of the total mass of the ISM, but this is still sufficient to block out much of the visible light.

The gas in the ISM is in a constantly changing chemical state. However, in the densest clouds most of the gas is normally molecular. This is firstly because the processes forming molecules in interstellar space – primarily two-body gas-phase reactions and catalysis on the surface of dust grains – proceed faster at higher density, and secondly

[†] 1 solar mass ($M_\odot$) = 2×10^{30} kg.
[‡] 1 parsec (pc) = 3×10^{16} m.

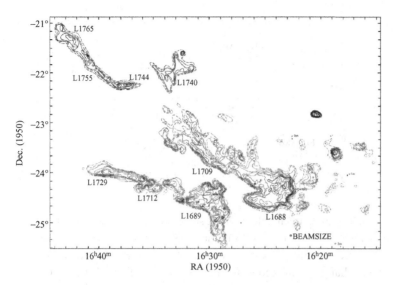

Fig. 1.3. A map of the Ophiuchus molecular cloud complex. The contours represent brightness of carbon monoxide (CO), which is taken as a tracer of the molecular gas as a whole.

because dust effectively shields the interior of a dense cloud from the ultraviolet (UV) radiation which destroys molecules.

Consequently the majority of the gas within the clouds is in molecular form, and the majority of a molecular cloud's mass is in the form of molecular hydrogen, H_2. Figure 1.3 shows a map of a molecular cloud in the constellation Ophiuchus. The sizes of molecular clouds range from ~ 0.1 pc to ~ 100 pc in diameter, and their densities range from 10^6 atoms per m^3 at their edges to greater than 10^{12} atoms per m^3 in their densest parts. GMCs are among the largest entities that we know of within galaxies.

In addition to gas and dust there are cosmic rays in the interstellar medium. Although these particles constitute a minute fraction of the rest mass in the interstellar medium, they travel at speeds approaching the speed of light, and consequently they have in total about as much kinetic energy as the rest of the matter put together.

The electromagnetic radiation in the interstellar medium comes from various sources. The principal sources of optical radiation are stars, but at longer wavelengths there is also continuum radiation from interstellar dust, line radiation from interstellar gas, and the cosmic microwave background (CMB – a relic of the Big Bang). Stellar radiation is dominated by optical and UV photons ($\lambda \sim 100$–1000 nm), corresponding to stellar surface temperatures in the range 3000–30 000 K. The cosmic microwave background is characterised by millimetre and submillimetre photons ($\lambda \sim 0.3$–3 mm $\equiv 3$–30×10^{-4} m). Infrared continuum radiation from dust falls in between the stellar radiation and the cosmic microwave background, with $\lambda \sim 30$–300 μm, corresponding to dust temperatures in the range $T_{dust} \sim 10$–100 K.

Fig. 1.4. Schematic view of the Galaxy from the side, showing the principal stellar components, the locations of the Galactic Centre and the Sun, and the location of the interstellar medium.

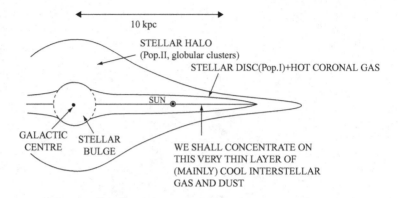

The gravitational field on large scales is the gravitational field of the Galaxy, which we believe to be dominated by stars and 'dark matter' (the precise nature of which is uncertain). Locally, the gravity of the interstellar gas may become important; this is particularly true in GMCs where the gas density is high and self-gravity may cause the interstellar gas to condense into new stars.

The interstellar medium is in a continual state of dynamical and chemical change, hence it is not in mechanical or chemical equilibrium. Additionally, the mean free paths for photons and cosmic rays in the interstellar medium are normally much longer than the typical distances over which the physical conditions change. Hence the interstellar medium is also not in local thermodynamic equilibrium (LTE). However, the mean free path of gas particles is short, so the gas is normally in thermal equilibrium, in the sense that there is a well-defined gas-kinetic temperature, T, characterising the distribution of gas particle velocities.

1.4 The distribution of the stars

We can identify several components of our Galaxy (the Milky Way).

If we could view it from the side we would see a distribution such as is shown in Figure 1.4. We start by describing how the stars are distributed. The oldest, most metal-poor stars[†] are known as Population II stars, and are distributed in a spheroidal (almost spherical) halo, which is at least 30 kpc across.[‡] This component includes the globular clusters.

There is also a roughly spherical bulge near the centre of the Milky Way, about 3 kpc in radius. The stars in the bulge are also old, but they have higher metallicity than the halo stars. Finally the youngest, most

---

[†] In astronomy, a metal is defined as any element other than hydrogen or helium. Any star with significantly fewer metals than the Sun is referred to as metal-poor. Any star with significantly more metals than the Sun is referred to as metal-rich.

[‡] 1 kpc $= 1000$ pc $= 3 \times 10^{19}$ m.

Fig. 1.5. The Orion Nebula, containing the Trapezium Cluster, as seen by the Hubble Space Telescope.

metal-rich stars (the Population I stars) are concentrated in a disc close to the midplane of the Milky Way. This stellar disc is at least 30 kpc across, and about 800 pc thick in the solar neighbourhood, although the more massive stars appear to be concentrated in a central layer only about 200 pc thick.

The Sun appears to be close to the midplane of the Milky Way (perhaps 10–20 pc above it), and about 8 kpc from the centre of the Milky Way. Most of the interstellar gas and dust is confined to an extremely thin layer, about 200 pc thick, inside the stellar disc. The radial extent of the gas disc is at least 20 kpc – i.e. at least 100 times its thickness – so the interstellar medium is like a very thin pancake.

There is some interstellar gas further from the midplane of the Milky Way. This is the coronal gas, very hot rarefied gas which is presumed to have escaped from the interiors of old supernova remnants, and which extends to about 3 kpc above and below the midplane. However, the total mass of the coronal gas is very small compared with the cooler interstellar gas near the Galactic midplane.

Stars and star clusters are observed to have a wide range of ages, from less than a million years for young stars like the Trapezium Cluster in Orion (see Figure 1.5), up to of order 10^{10} years for the oldest globular clusters. This implies that star formation has occurred since very early times – the age of the Universe is estimated to be $\sim 1.3 \pm 0.3 \times 10^{10}$ years – and is still continuing.

Star formation converts diffuse interstellar gas clouds into star clusters. The rate at which interstellar gas is converted into stars has a profound effect on the overall dynamics of a galaxy, and hence on its

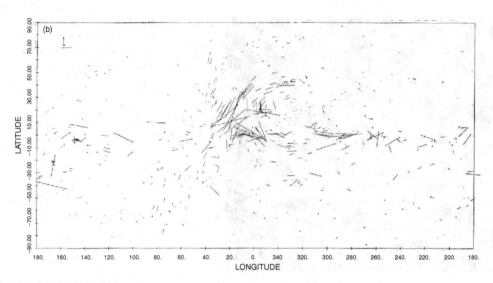

Fig. 1.6. The polarisation of starlight projected onto Galactic coordinates. This is believed to be tracing the large-scale magnetic field in the interstellar medium of our Galaxy.

formation, structure and evolution. Thus, the star formation within a galaxy is crucial to that galaxy's overall evolution.

1.5 The magnetic field

The origin of the interstellar magnetic field is uncertain. Usually it is assumed to be generated by a galactic-scale dynamo, and amplified locally by gas-dynamical processes, but the details of the underlying mechanisms are not well understood. The strength of the magnetic field is typically $|\mathbf{B}| \sim 3$ μgauss[†] and so its energy density is $|\mathbf{B}|^2/4\pi \sim 10^{-13}$ J m^{-3}, which is comparable with the energy densities of the gas, of the radiation field, and of the cosmic rays. Therefore the magnetic field is likely to play an important role in determining the structure and evolution of the interstellar medium.

A number of observations lead astronomers to believe that magnetic fields are ubiquitous in the ISM. This is not a totally surprising result given that magnetic fields are caused by moving charges and that the ISM appears to be in constant motion. One of the main pieces of evidence we have for magnetic fields comes from observations of the polarisation of starlight. When such polarisation measurements were first carried out, astronomers were surprised to find that the light from stars was partially plane polarised in a pattern across the sky (see Figure 1.6). Furthermore, this pattern appears to trace out large-scale structures along the plane of the Galaxy.

If the polarisation were a property purely of the stars themselves it would be very difficult to explain this apparent structure. However,

[†] 1 gauss $= 10^{-4}$ tesla, so 1 μgauss $= 10^{-10}$ tesla.

astronomers believe that the polarisation is not a property of the stars themselves but of the ISM between the stars and us, and that it is caused by an interstellar magnetic field.

To understand how this mechanism works we must realise that interstellar dust grains are not spherically symmetrical, but typically have very complex shapes. We usually try to simplify our consideration of their shapes by thinking of them either as needle-like cylinders or prolate spheroids (rugby ball or American football shaped). In this way we can think of their asymmetry simply in terms of a 'long axis' and a 'short axis'.

Put simply, the long axis of a grain extinguishes the background starlight more efficiently than the short axis. Hence if a large ensemble of non-spherical dust grains is aligned preferentially in one direction, the background starlight is preferentially extinguished along one axis, causing the transmitted light to be partially plane polarised.

The manner of the alignment is still a matter of debate, and many theories have been put forward to explain it. However, most theories predict that the grains spin around the direction of the magnetic field, with their long axis perpendicular to the field. The original version of this mechanism is known as the Davis–Greenstein effect. The basis of this effect is as follows:

The dust grains in a molecular cloud are in random motion and undergo frequent collisions. These collisions set the grains spinning. The grains are typically composed of silicate material which is paramagnetic in nature. If the cloud is threaded by a magnetic field, then the presence of this external field causes an induced internal field within the paramagnetic dust grain material, whose strength depends on the magnetic susceptibility of the material. Normally these two fields would be parallel to one another, but because the grain is spinning, the internal field cannot respond quickly enough, so it always 'lags' behind the external field direction. This causes a net torque which tends to cause the grain to spin with its long axis perpendicular to the external magnetic field direction. Further collisions of course serve to misalign the grains once more.

This mechanism may not be exactly correct in practice as it requires a field strength about an order of magnitude higher than that which is measured (see below). However, a number of alternatives have been proposed. One plausible mechanism is super-paramagnetic alignment, which requires inclusions of a ferromagnetic substance within the dust grain material to increase the alignment efficiency. Another possibility is suprathermal alignment in which molecules being ejected from the grain surface help to spin up the grain and hence shorten the alignment time-scale and increase the mechanism's efficiency. Whichever of these mechanisms ultimately proves correct, all agree that the grains become preferentially aligned perpendicular to the magnetic field direction. Hence the background starlight is preferentially extinguished in this direction. The transmitted light is therefore partially plane polarised parallel to the magnetic field direction.

Linear polarisation allows us to measure the orientation of the magnetic field in the plane of the sky. To measure the strength of the magnetic field we use the Zeeman effect, which relies on the splitting of degenerate atomic or molecular energy levels in the presence of a magnetic field. The amount of the splitting is proportional to the field strength, thus allowing the field strength to be measured.

Consider the simplest case of a hydrogen atom. The electronic ground state (s state) has principal quantum number $n = 1$ and angular momentum quantum number $l = 0$. The first excited state (p state) has $n = 2$ and $l = 1$. The magnetic quantum number m_l must obey the relation $|m_l| \leq l$. Hence the allowed values for m_l in the p state are $0, +1$ and -1, and the state is said to have triple degeneracy. In the presence of a magnetic field the degeneracy is lifted and the p state becomes a triplet. Hence the spectral line of the transition between p and s states becomes a triplet. Each level is shifted in energy E by an amount

$$E = \mu_B m_l B, \tag{1.1}$$

where the constant μ_B is known as the Bohr magneton and has a value of 9.27×10^{-24} J/T, and B is the magnitude of the magnetic field strength along the line of sight to the observer.

Hence by observing such a multiple line whose values of m_l are known, we can measure the magnetic field strength in one direction. Typical values that have been measured in the ISM are $\sim 10^{-10}$–10^{-9} T. These values are so small that the typical splitting is too small for most spectrometers to measure. However, the two levels have opposite circular polarisations and can therefore be split by a polarimeter, which is sensitive to circular polarisation. This means therefore that only the field strength along the line of sight is measured and various geometric assumptions have to be made to infer the three-dimensional magnetic field configuration. In Chapter 4 we will discuss the effects that magnetic fields have on the dynamics of the ISM.

1.6 Star formation in a galactic context

Star formation converts diffuse interstellar gas clouds, which undergo highly dissipative collisions and are therefore very inelastic, into star clusters, which are effectively collisionless and therefore much more elastic. What this means is that when two clouds of interstellar gas collide, the mean free path for collisions between the individual gas

Fig. 1.7. Infrared image of the spiral galaxy M81 showing the star-formation regions strung out along the spiral arms of the galaxy. Inset is an optical image illustrating that these regions are much less clear in the optical. This is because optical wavelengths are extinguished by the dust in star-formation regions.

particles is very short. Consequently, the bulk kinetic energy of the clouds is converted into random thermal energy of the gas particles and then radiated away – i.e. the gas is first heated by the collision, and then it cools by emitting radiation. Therefore, the bulk kinetic energy of the clouds is lost irreversibly and entropy is created. The collision between the two clouds is extremely inelastic and dissipative.

By contrast, when two clouds of stars – i.e. two star clusters – collide, the mean free path for collisions between individual stars is extremely long, in fact there is virtually no chance that two stars will collide. Therefore a collision between two clouds of stars is non-dissipative and elastic. We should distinguish two limiting cases.

(i) If there is only a small number of stars in each cloud, and they collide at high speed, the two clouds pass straight through one another.

(ii) Conversely, if there is a larger number of stars in each cloud, and the clouds collide at lower speed, long-range gravitational interactions between the stars produce significant deflections of their individual orbits. In this limit, the bulk kinetic energies of the two clouds can be randomised, and the result is a single merged cloud of stars. This merged cloud is more extended and cooler than either of the original clouds, in the sense that the random motions of the individual stars are slower than they were in the original clouds. However, no energy has been radiated away; it is still all invested in the kinematics of individual stars.

Figure 1.7 shows an infrared image of the spiral galaxy M81. The spiral structure of the galaxy can be clearly seen, as can the bright

Fig. 1.8. An optical image of the globular cluster M80.

regions of star formation delineating the spiral arms. Inset is an optical image of the same galaxy, in which the spiral structure can still be seen, but the star-formation regions are less clear. This is because the optical emission is more efficiently extinguished than the infrared by the dust in the star-formation regions.

1.7 Known sites of contemporary star formation

Stars form in molecular clouds, which are concentrated in the discs of galaxies like the Milky Way, particularly in their spiral arms; in irregular galaxies, like the Magellanic Clouds; in starburst galaxies, like M82; and in interacting and merging galaxies, like the Antennae. In the Milky Way, there are modest star-formation regions, like Taurus, and more prolific ones, like Orion, W3 and W49. In external galaxies there are even more vigorous star-formation regions like 30 Doradus in the Large Magellanic Cloud, and the active galaxy Arp 220.

Most stars – and possibly all stars – are born in clusters[†] (see Figure 1.8), although these clusters may involve as few as 10 members or as many as 10^7. Some clusters are relatively small and nearby, such as the Pleiades Cluster (see Figure 1.9). On a larger scale, violently interacting galaxies trigger the formation of very large clusters, reminiscent of the old globular clusters found in the halo of the Milky Way, with 10^5–10^6 stars and diameter ~ 40 pc (see Figure 1.8).

[†] We shall use the term cluster to mean any collection of stars whose formation appears to have been related.

Fig. 1.9. An optical image of the Pleiades open cluster.

In the neighbourhood of the Sun, the largest coherent clusters of young stars are called OB associations. OB associations are so called because they contain significant numbers of massive O and B stars, and they are therefore also the locations of extended HII regions. These are regions of ionised gas around massive stars (HII is singly ionised hydrogen).

A typical OB association contains between 3000 and 10^5 stars, and has a diameter between 10 and 200 pc. OB associations are unbound, so their diameter increases with age. By the time the O and B stars have burnt out, after a few tens of Myrs, an association is largely dispersed, and hard to discern. OB associations represent the aftermath of star formation in a large molecular cloud complex, which is then dispersed by the action of the O and B stars. Nearby examples include Orion, Upper Scorpius and Upper Centaurus–Lupus.

Within an OB association, there may be several subgroups and/or embedded clusters, representing local regions of star formation at different stages of evolution. Embedded clusters are the youngest regions of star formation, where the stars are still surrounded by the residual gas and dust from which they formed, and therefore they can only be observed at infrared wavelengths; a good example is the embedded cluster NGC 2024 in the Orion B molecular cloud (see Figure 1.10). Subgroups are optically visible clusters, which have dispersed most of the residual gas and dust from which they formed. There is evidence for self-propagating star formation in many OB associations – see Figure 1.11.

Fig. 1.10. Sequential, self-propagating star formation, as seen in the Orion region (see also Figure 1.5). The solid contours show the molecular cloud. The dashed contours encircle the main OB association subgroups, which are labelled as 1a, 1b and 1c. Orion 1a is the oldest, 1b is the next oldest, and 1c is the youngest. Compare this with the theoretical picture in Figure 1.11.

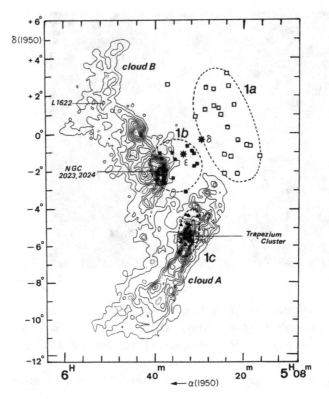

This is presumed to occur when the massive stars of one subgroup excite HII regions, blow stellar winds, and finally explode as supernovae. The resulting expanding nebulae compress the surrounding gas, thereby triggering the formation of the next subgroup. This process can repeat itself recursively, thereby generating a spatial and temporal sequence of ever younger subgroups. A nearby example is the sequence of subgroups terminating at the Trapezium Cluster in Orion; in order of decreasing age, these subgroups are Orion 1a, Orion 1b and Orion 1c. This is shown in Figure 1.10.

Most subgroups are unbound from the outset, but a few are initially bound. These are the open (or Galactic) clusters. Typically they contain between 30 and 1000 stars, and have diameters between 1 and 20 pc. They normally survive for several crossing times, but eventually they too are dispersed, due to evaporation and/or the tidal perturbations of passing clouds. Few last beyond about 10^9 years. Famous examples are Praesepe, the Hyades and the Pleiades (see Figure 1.9).

Finally there are T associations containing 30–300 stars, in a region of diameter 3–30 pc. T associations are so called because they contain

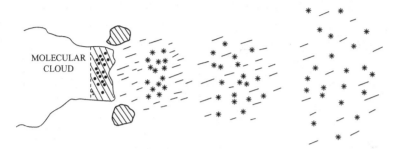

Fig. 1.11. Theoretical, sequential, self-propagating star formation, driven by the expansion of an HII region.

no O or B stars, but only much lower-mass T Tauri stars. The archetype is the Taurus–Auriga T association, at a distance of $\sim$140 pc.

1.8 The initial mass function

We define the initial mass function (IMF) for star formation, $\phi(M)$, such that, if a net mass ΔS is converted into new stars, the number of stars in the mass interval $(M, M + dM)$ is given by

$$\mathcal{N}_M \, dM = \Delta S \, \phi(M) \, dM. \tag{1.2}$$

$\phi(M)$ is normalised such that

$$\int_{M_{\min}}^{M_{\max}} \phi(M) \, M \, dM = 1. \tag{1.3}$$

The first person to derive an IMF was Salpeter. In 1955, he fitted the available observational data with a power law between $0.4 \, M_\odot$ and $10 \, M_\odot$. He found that

$$\phi(M) \, dM \simeq K \, M^{-2.35} \, dM, \tag{1.4}$$

where K is a constant.

Today the IMF can be measured, both in clusters, and in the field, to below the hydrogen-burning limit at $\sim$0.075 $M_\odot$, and significant departures from equation 1.4 are found in this limit. The IMF is usually fitted with piece-wise power laws. For example, one commonly accepted form for the IMF is

$$\phi(M) \, dM \simeq K \, M^{-2.3} \, dM : M \stackrel{>}{\sim} 0.5 M_\odot \tag{1.5}$$

$$\phi(M) \, dM \simeq K \, M^{-1.3} \, dM : 0.5 M_\odot \stackrel{>}{\sim} M \stackrel{>}{\sim} 0.08 M_\odot \tag{1.6}$$

$$\phi(M) \, dM \simeq K \, M^{-0.3} \, dM : 0.08 M_\odot \stackrel{>}{\sim} M \stackrel{>}{\sim} 0.01 M_\odot, \tag{1.7}$$

where we note that the Salpeter-like form still holds for higher masses. This form of the IMF is shown in Figure 1.12.

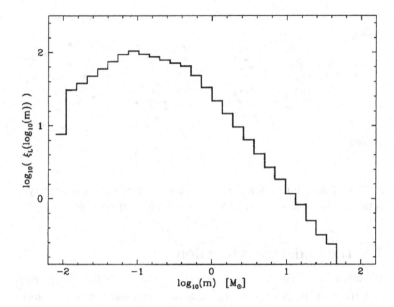

Fig. 1.12. The initial mass function, here plotted as $\log(\xi)$, where $\xi = M\phi(M)$, against $\log(m)$, where $m = M/M_\odot$.

The IMF can also be fitted with a log-normal distribution. For example, recent work finds

$$\phi(M)dM \simeq 0.584 M_\odot^{-1} \exp\left\{-1.54 \log^2\left[\frac{M}{0.22 M_\odot}\right]\right\}\frac{dM}{M} \qquad (1.8)$$

where we have normalised in accordance with equation 1.3.

One of the goals of star formation is to understand why the stellar IMF appears to vary very little from one star-formation region to another, and to explain any variations in terms of environmental effects. We now list some of the main goals for any star-formation theory.

1.9 Objectives of star-formation theory

The overall goal of star-formation research is to develop a complete theory for the whole process. However, we are far from this overall goal, so it is appropriate to define some more specific objectives, in the form of questions. We list these questions chronologically, in the order that the events implicit in the questions are presumed to occur, rather than according to any hierarchy of importance.

1. Is there a threshold for star formation to occur? For instance, it has been suggested that a galactic disc cannot fragment unless

$$Q \equiv \frac{a_0 \kappa}{\pi G \Sigma} < 1, \qquad (1.9)$$

and hence that this is a threshold for star formation in a galactic disc. Here Q is known as the Toomre parameter, a_0 is the effective sound speed in the disc, κ is the epicyclic frequency in the disc, or the frequency of oscillation of matter about its mean orbital radius, G is the gravitational constant and Σ is the surface density of the disc. There is some observational evidence to support this idea, but it is certainly only part of the full star-formation story. On a smaller scale it has been suggested that within individual molecular clouds (see Chapter 4) there is a density threshold below which star formation cannot occur. This has been explained theoretically in terms of external ionisation of a molecular cloud and subsequent magnetically regulated collapse. We return to this in Chapter 5.

2. If there is a threshold for star formation, what is the efficiency of star formation once this threshold is passed? We define the star-formation efficiency η_{SF} as

$$\eta_{SF} = \frac{\Delta S}{\mathcal{G}_0}, \tag{1.10}$$

where $\mathcal{G}_0$ is the total initial mass of interstellar matter involved, and ΔS is the mass converted into new stars. The answer to this question is related to the detailed mapping between the core mass function and the stellar initial mass function discussed in Chapter 5.

3. If there is no threshold, then what determines the mean rate of star formation? We define the star-formation rate $\bar{\mathcal{R}}_{SF}$ as

$$\bar{\mathcal{R}}_{SF} \equiv \left\langle \frac{1}{\mathcal{G}} \frac{dS}{dt} \right\rangle, \tag{1.11}$$

where $\mathcal{G}$ is the current total mass of interstellar matter, and dS/dt is the current rate at which that matter is being converted into stars. We look at this in the context of Galaxy-wide star formation in Chapter 8.

4. What causes the initial mass function (IMF) for star formation, $\phi(M)$? One interesting aspect of the IMF is that it appears to differ very little from one region to another, and therefore it may be essentially universal. If so, then this must be telling us something fundamental about the star-formation process. In Chapter 5 we discuss this in terms of the mass function of the cores from which stars are formed.

5. What fraction of all stars is born in clusters, as opposed to being born in isolation? The dividing line between a very sparse cluster and truly isolated star formation may be hard to define precisely. This question cannot presently be answered by theory alone. Detailed observations are required. Current thinking is that most stars are formed in clusters, but the exact percentage remains unclear.

6. When stars are born in a cluster, what factors determine whether it is a massive tightly bound globular cluster, a medium-mass loosely bound open cluster, or a low-mass unbound association? This question is related to the star-formation efficiency question above, and is a matter of ongoing research.

7. What determines the statistics of binary systems? By this we mean the distributions of mass ratio, period, and eccentricity as a function of the mass of the primary star.

Concentrating first on Sun-like primaries, with masses in the range of $\sim$0.5–2 $M_\odot$, the following facts must be explained: (i) about 60% of Sun-like field stars[†] are in binaries or higher hierarchical multiples; (ii) the components in Sun-like field binary systems appear to pick their partners more-or-less at random as regards mass; (iii) there is a wide distribution of periods from less than an hour to nearly 10^5 years with a peak around 200 years; and (iv) there is a wide range of eccentricities, except for the very short-period systems ($P \stackrel{<}{\sim} 10$ days) which almost always have circular orbits, due to tidal circularisation.

Turning now to the full range of primary star masses, it appears that, as the primary mass decreases, the binary fraction decreases, the mean mass ratio increases (tending towards approximately equal-mass components for brown dwarf binaries), and the mean period and range of periods both decrease.

Furthermore, there is a growing body of evidence which suggests that newly formed stars have an even higher binary fraction than field stars. In fact, conceivably, *all* stars may be formed in binaries or higher multiples, and then some of these binaries are subsequently disrupted to produce singles. Thus the formation of binaries appears to be an integral part of the formation of stars. We return to the topic of binary stars in Chapter 5.

8. Do most stars have planetary systems? The formation of our planetary system appears to have occurred at the same time as the formation of the Sun. The material which went into the Sun is presumed to have done so by transferring most of its angular momentum to a circumsolar disc of gas and dust. The planets then condensed out of this disc. Discs with the appropriate dimensions and masses have been seen around many young stars. A related question is what fraction of planetary systems include Earth-like planets? This topic is related to the physics of circumstellar discs. We discuss this further in Chapter 8.

9. How do all these various aspects of star formation depend on environmental factors? These may vary enormously with epoch. By

---

[†] By 'field stars' we mean stars not currently associated with a given cluster or association.

'environmental factors' we mean external factors such as: (i) mergers of galaxies and close interactions between galaxies; (ii) the strength of the shock waves associated with spiral modes in galaxies, and the depth of the galactic potential well in which the star formation occurs; (iii) the level of turbulence in the interstellar medium, and the strength of the interstellar magnetic field; (iv) the ambient intensity of high-energy ionising radiation such as UV and X-ray photons, and cosmic rays; (v) the temperature of the cosmic microwave background radiation field; and (vi) the heavy-element abundance in the interstellar gas. This latter abundance is usually referred to as the 'metallicity', Z, expressed as the fraction of the total mass in elements other than H or He. We discuss galaxy mergers in Chapter 8. Turbulence, magnetic fields and ionisation are discussed in Chapters 4 and 5. The star-formation rate as a function of epoch, metallicity and microwave background variations is discussed in Chapters 7 and 8.

10. What implications does star formation have for other areas of research? Specifically, these include: (i) the formation, structure and evolution of galaxies; (ii) cosmochemistry (the origin and distribution of the chemical elements) – heavy elements are both formed in stars, and locked up in low-mass or dead stars, and so the yield of heavy elements from a generation of stars depends critically on its initial mass function; (iii) the existence of life elsewhere in the Universe is dependent on whether there are other Earth-like planets out there; since planet formation appears to be a by-product of star formation, the answer to this question is intimately linked to star-formation theory; (iv) the heat-death of the Universe – since stars are the most ubiquitous and efficient manufacturers of entropy in the Universe, star formation controls the rate of heat-death. We return to some of the consequences of star formation in Chapter 8.

The reader should be aware that none of these questions has a definitive answer. This field is one of continuing research and discovery. However, in this book we hope to give the reader some idea of how these questions are being addressed.

Recommended further reading

We recommend the following texts to the student for further study.

Adamson, A., *et al.* (2005). *Astronomical Polarimetry: Current Status and Future Directions*. Astronomical Society of the Pacific Conference Series, vol. 343. San Francisco: Astronomical Society of the Pacific.

Bally, J. and Reipurth, B. (2006). *The Birth of Stars and Planets*. Cambridge: Cambridge University Press.

Corbelli, E., Palla, F. and Zinnecker, H. (2005). *The Initial Mass Function Fifty Years Later*. Astrophysics and Space Science Library, vol. 327. Dordrecht: Springer.

Kroupa, P. (2002). The initial mass function of stars. *Science*, vol. 295, pp. 82–91.

Smith, M. D. (2004). *The Origin of the Stars*. London: Imperial College Press.

Stahler, S. W. and Palla, F. (2004). *The Formation of Stars*. Weinheim: Wiley-VCH.

Tayler, R. J. (1981). *The Stars: Their Structure and Evolution*. London: Wykeham Publications.

Chapter 2
Probing star formation

2.1 Introduction

In the preceding chapter we discussed the main constituents of a galaxy. In this chapter we describe the ways in which we detect and measure those constituents. We also discuss the chief component that we have not yet mentioned – the radiation field.

We describe the various ways in which we learn about the Universe. We introduce fundamental concepts such as intensity, flux and opacity, and we show how these can be applied to both continuum radiation and spectral line radiation. These ideas are then used to illustrate how we can learn about the physical properties of the gas and dust in the interstellar medium.

2.2 Properties of photons

The majority of what we know about the Universe comes as a direct result of the electromagnetic (EM) radiation we receive from the Universe.[†]

The other ways in which we learn about the Universe are space probes that travel to other bodies in the Solar System to discover the details of their composition, and from the meteors and meteorites that fall to Earth from time to time. Space probes can help us to learn about the formation of our own star, the Sun, and its planets, from which we may be able to extrapolate to the formation of other stars and their planets. The study of meteors can tell us about inter-planetary dust,

---

[†] Other potential sources of information include cosmic rays, neutrinos and gravitational waves, but these are generally of little use in studies of star formation.

Fig. 2.1. The spatial and
directional dependence of the
intensity of a radiation field.

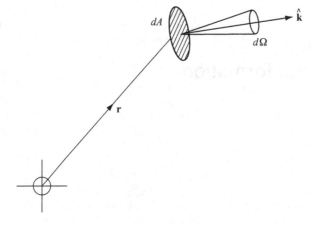

but it is by the study of EM radiation that we learn the most about the Universe.

EM radiation consists of individual quantised particles known as photons. Each photon has only four properties: the direction from which it emanates; the time of its arrival; its polarisation; and its associated energy E.[†] The energy of a photon is related to its frequency v by the simple relation $E = hv$ and to its wavelength λ by the equation $E = hc/\lambda$, where c is the speed of light in a vacuum and h is Planck's constant. The direction of travel of a photon helps us to determine from which region of space it originated; the time of arrival is used in the study of variability of stars; polarisation can give us information on magnetic fields; and the energies of the photons determine the conditions in the object being studied.

When we have large numbers of photons we can study the statistical properties of this radiation. We will distinguish between two types of radiation, which we will refer to as 'continuum' and 'line' radiation. Continuum radiation is distributed over a wide range of frequencies. Line radiation peaks at specific frequencies. The study of continuum radiation is known as photometry, and the study of line radiation is referred to as spectroscopy.

2.3 Intensity

A steady unpolarised radiation field is completely described by the monochromatic intensity I_v. Intensity is a useful quantity, as in a vacuum it is conserved along a ray. I_v gives the amount of radiant energy in unit

---

[†] Our knowledge of both the energy and arrival time simultaneously are of course limited by the uncertainty principle.

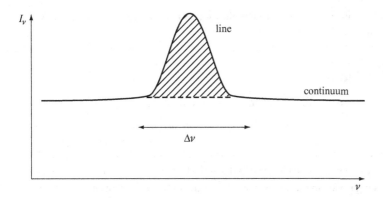

Fig. 2.2. The integrated intensity of a line is obtained by subtracting the underlying continuum intensity; this leaves the shaded area.

frequency interval about ν, crossing unit area, in unit time, into unit solid angle about the normal to that area. Alternatively,

$$I_\nu(\mathbf{r}, \hat{\mathbf{k}}) \, d\nu \, dA(\mathbf{r}, \hat{\mathbf{k}}) \, dt \, d\Omega(\hat{\mathbf{k}}) \qquad (2.1)$$

gives the amount of radiant energy in the infinitesimal frequency interval $(\nu, \nu + d\nu)$, passing through the infinitesimal element of area $dA(\mathbf{r}, \hat{\mathbf{k}})$ having position $\mathbf{r}$ and unit normal $\hat{\mathbf{k}}$, into the infinitesimal element of solid angle $d\Omega(\hat{\mathbf{k}})$ about direction $\hat{\mathbf{k}}$, during the infinitesimal time interval $(t, t + dt)$ – see Figure 2.1.

Here 'monochromatic' means that I_ν describes the intensity at 'one colour', i.e. a single frequency ν. Remember that the frequency interval $d\nu$ is infinitesimal, so by implication $d\nu \to 0$. The *monochromatic* intensity I_ν must be distinguished from the *integrated* intensity I (with no subscript ν). I is given by

$$I(\mathbf{r}, \hat{\mathbf{k}}) = \int_{\nu=0}^{\nu=\infty} I_\nu(\mathbf{r}, \hat{\mathbf{k}}) \, d\nu, \qquad (2.2)$$

and measures the amount of radiant energy, summed over all frequencies, passing through unit area at position $\mathbf{r}$, in unit time, into unit solid angle about the unit normal $\hat{\mathbf{k}}$ to that area.

When we speak about the integrated intensity of a spectral line, we subtract the underlying continuum intensity before performing the integration over all frequencies (see Figure 2.2). Additionally, the limits of integration are reduced to avoid contamination by other lines:

$$I^{\text{line}}(\mathbf{r}, \hat{\mathbf{k}}) = \int_{\nu=\nu_0-\Delta\nu/2}^{\nu=\nu_0+\Delta\nu/2} \left[I_\nu^{\text{observed}}(\mathbf{r}, \hat{\mathbf{k}}) - I_\nu^{\text{continuum}}(\mathbf{r}, \hat{\mathbf{k}}) \right] \, d\nu. \qquad (2.3)$$

Here ν_0 is the frequency at the line centre, $\Delta\nu$ is the linewidth, I_ν^{observed} is the observed intensity, and $I_\nu^{\text{continuum}}$ is the background continuum.

$I_\nu^{\text{continuum}}$ is estimated by interpolation, as indicated by the dashed line in Figure 2.2.

2.4 Flux

We can also define the monochromatic flux, F_ν, such that

$$F_\nu(\mathbf{r}, \hat{\mathbf{k}})\, d\nu\, dA(\mathbf{r}, \hat{\mathbf{k}})\, dt \tag{2.4}$$

gives the *net* amount of radiant energy in the infinitesimal frequency interval $(\nu, \nu + d\nu)$, crossing the infinitesimal element of area $dA(\mathbf{r}, \hat{\mathbf{k}})$ having position $\mathbf{r}$ and unit normal $\hat{\mathbf{k}}$, during the infinitesimal time interval $(t, t + dt)$, *irrespective* of the direction of the radiation relative to $\hat{\mathbf{k}}$.[†]

The integrated flux is then given by

$$F(\mathbf{r}, \hat{\mathbf{k}}) = \int_{\nu=0}^{\nu=\infty} F_\nu(\mathbf{r}, \hat{\mathbf{k}})\, d\nu. \tag{2.5}$$

Strictly speaking, the monochromatic and integrated fluxes are vector fields, $\mathbf{F}_\nu(\mathbf{r})$ and $\mathbf{F}(\mathbf{r})$, which give the direction in which the flow of energy is a maximum. The scalar fluxes we have defined are then related to the vector fluxes by

$$F_\nu(\mathbf{r}, \hat{\mathbf{k}}) = \mathbf{F}_\nu(\mathbf{r}) \cdot \hat{\mathbf{k}}, \qquad F(\mathbf{r}, \hat{\mathbf{k}}) = \mathbf{F}(\mathbf{r}) \cdot \hat{\mathbf{k}}. \tag{2.6}$$

The relation between flux and intensity is

$$F_\nu(\mathbf{r}, \hat{\mathbf{k}}) = \int I_\nu(\mathbf{r}, \hat{\mathbf{k}}')\, \hat{\mathbf{k}}' . \hat{\mathbf{k}}\, d\Omega(\hat{\mathbf{k}}'), \tag{2.7}$$

where we have introduced a second unit vector $\hat{\mathbf{k}}'$ to denote the direction of the intensity (see Figure 2.1).

2.5 Radiant energy density

The monochromatic radiant energy density, u_ν, is defined so that

$$u_\nu(\mathbf{r})\, d\nu\, dV(\mathbf{r}) \tag{2.8}$$

gives the amount of radiant energy in the infinitesimal frequency interval $(\nu, \nu + d\nu)$, in the infinitesimal volume element $dV(\mathbf{r})$ at position $\mathbf{r}$.

The integrated radiant energy density is then given by

$$u(\mathbf{r}) = \int_{\nu=0}^{\nu=\infty} u_\nu(\mathbf{r})\, d\nu. \tag{2.9}$$

The monochromatic volume emissivity, j_ν, is defined so that

$$j_\nu(\mathbf{r})\, dV(\mathbf{r})\, dt\, d\Omega(\hat{\mathbf{k}})$$

[†] Note how this definition differs from the definition of the intensity.

gives the amount of radiant energy emitted from the infinitesimal volume element $dV(\mathbf{r})$ at position $\mathbf{r}$, during the infinitesimal time interval $(t, t + dt)$, into the infinitesimal element of solid angle $d\Omega(\hat{\mathbf{k}})$ about the direction $\hat{\mathbf{k}}$. We are implicitly assuming that the emission is isotropic, so j_ν does not depend on $\hat{\mathbf{k}}$.

The integrated volume emissivity, $j(\mathbf{r})$, is given by

$$j(\mathbf{r}) = \int_{\nu=0}^{\nu=\infty} j_\nu(\mathbf{r}) \, d\nu. \tag{2.10}$$

The monochromatic volume opacity, κ_ν, is the total effective absorption cross-section presented by all the matter in unit volume to photons of frequency ν. We say effective because it is corrected to include stimulated emission (which is equivalent to negative absorption). κ_ν is the same as the inverse of the mean free path l_ν for photons of frequency ν. Hence,

$$\kappa_\nu(\mathbf{r}) = \sum_X \{n_X(\mathbf{r}) \sigma_X(\nu)\} = l_\nu^{-1}(\mathbf{r}). \tag{2.11}$$

Here $n_X(\mathbf{r})$ is the number density of particles of species X at position $\mathbf{r}$, $\sigma_X(\nu)$ is the cross-section presented by a single particle of species X to radiation of frequency ν, and the sum is over all species X. At certain frequencies the sum is dominated by the contribution from a single species, which has a strong absorption cross-section at that frequency. For stimulated emission the effective cross-section is negative.

2.6 Continuum radiation – studying the dust

Interstellar dust consists of very small grains, less than a micron in size, that are mostly silicates and carbon compounds (see Section 1.3). By mass, the dust represents about 1% of the total mass of the ISM. It is possible to study the emission from dust by observing at particular wavelengths.

Dust in the ISM emits radiation over a broad range of wavelengths, and in the range from ~ 10 μm to $\sim 10^3$ μm ($= 1$ mm) dust emission dominates the radiation from the ISM. In order to analyse the emission from interstellar dust, we first introduce the notion of blackbody radiation.

A blackbody radiation field is a uniform and isotropic radiation field in thermodynamic equilibrium. The monochromatic intensity of a blackbody radiation field is given by the Planck function, $B_\nu(T)$, where

$$I_\nu(\mathbf{r}, \hat{\mathbf{k}}) = B_\nu(T) = \frac{2h\nu^3}{c^2} \left\{ \exp\left[\frac{h\nu}{kT}\right] - 1 \right\}^{-1}, \tag{2.12}$$

where k is the Boltzmann constant and h is the Planck constant.

The monochromatic energy density of a blackbody radiation field is then given by

$$u_\nu(\mathbf{r}) = \frac{4\pi B_\nu(T)}{c}, \tag{2.13}$$

and the monochromatic flux is zero ($F_\nu = 0$) because a blackbody radiation field is isotropic.

The integrated intensity of a blackbody radiation field is given by

$$I = \frac{\sigma_{SB} T^4}{\pi}, \tag{2.14}$$

where σ_{SB} is the Stefan–Boltzmann constant

$$\sigma_{SB} = \frac{2\pi^5 k^4}{15c^2 h^3} = 5.7 \times 10^{-8} \, \mathrm{J\,m^{-2}\,s^{-1}\,K^{-4}}. \tag{2.15}$$

The integrated energy density is

$$u = \mathbf{a} T^4, \tag{2.16}$$

where $\mathbf{a}$ is a constant equal to

$$\mathbf{a} = \frac{8\pi^5 k^4}{15(hc)^3} = 7.6 \times 10^{-16} \, \mathrm{J\,m^{-3}\,K^{-4}}. \tag{2.17}$$

The integrated flux is zero ($F = 0$) because the integrated intensity is isotropic.

When we speak of a stellar surface radiating like a blackbody at temperature T_*, we mean that immediately above the surface of the star the intensity over outward directions is like a blackbody radiation field at temperature T_*, and the intensity over inward directions is negligible. Hence the monochromatic and integrated fluxes are given by

$$\mathbf{F}_\nu(\mathbf{r}) \simeq \pi B_\nu(T_*) \, \hat{\mathbf{r}}, \tag{2.18}$$

and

$$\mathbf{F}(\mathbf{r}) \simeq \sigma_{SB} T_*^4 \, \hat{\mathbf{r}}, \tag{2.19}$$

where $\hat{\mathbf{r}}$ is the unit radial vector.

Blackbody radiation has one other relevant property, which relates the temperature, T, of the blackbody to the wavelength at which its emission peaks, λ_{max}, according to the equation

$$\lambda_{max} T = 2.898 \times 10^{-3} \, \mathrm{K\,m}. \tag{2.20}$$

This is known as Wien's displacement law. Note that the units of the constant on the right-hand side are kelvins × metres.

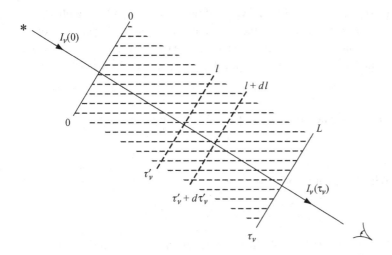

Fig. 2.3. The line of sight to a background source through an optically active medium.

2.7 Radiative transfer

The manner in which radiation interacts with the medium through which it travels on its way to an observer is known as radiative transfer. We consider transfer of radiation through a medium which extends from $l = 0$ to $l = L$ along the direction of propagation (see Figure 2.3). Radiation enters the medium at $l = 0$ with intensity $I_\nu(0)$, and emerges at $l = L$ with intensity $I_\nu(L)$. $I_\nu(0)$ is called the background intensity. $I_\nu(L)$ is what the observer sees.

The equation describing the change in the intensity of radiation as it passes through a medium is

$$\frac{dI_\nu}{dl}(l) = -\kappa_\nu(l)\, I_\nu(l) + j_\nu(l). \tag{2.21}$$

The first term on the right-hand side represents *attenuation of the intensity* due to the effective absorption (the difference between true absorption and stimulated emission). The second term represents the *increase in intensity* due to spontaneous emission in the medium – see equations 2.10 and 2.11.

The reason why the stimulated emission is included with the true absorption is that both processes have rates (per unit volume) which depend on the incident intensity. By contrast, the rate of spontaneous emission (per unit volume) is independent of the incident intensity.

Equation 2.21 can be simplified if we introduce two new quantities. The first is the optical depth τ'_ν, which is defined by

$$d\tau'_\nu(l) = \kappa_\nu(l)\, dl, \tag{2.22}$$

$$\implies \tau'_\nu(l) = \int_{l'=0}^{l'=l} \kappa_\nu(l')\, dl'. \tag{2.23}$$

If we recall that κ_ν is the inverse of the mean free path for a photon of frequency ν, we see that the optical depth measures distances through the medium in units of the local mean free path. Hence the optical depth is dimensionless. The total optical depth through the medium is

$$\tau_\nu = \tau'_\nu(L). \tag{2.24}$$

The second new quantity is the source function, S_ν, where

$$S_\nu = \frac{j_\nu}{\kappa_\nu}. \tag{2.25}$$

With these definitions, equation 2.21 reduces to

$$\frac{dI_\nu}{d\tau'_\nu}\left(\tau'_\nu\right) = -I_\nu\left(\tau'_\nu\right) + S_\nu\left(\tau'_\nu\right). \tag{2.26}$$

We note parenthetically that in thermodynamic equilibrium (TE) the intensity is uniform $(dI_\nu/d\tau'_\nu = 0)$ and given by the Planck function $B_\nu(T)$, so

$$S_\nu = I_\nu = B_\nu(T) \quad \text{(in TE)}, \tag{2.27}$$

and

$$j_\nu = \kappa_\nu\, B_\nu(T) \quad \text{(in TE)}. \tag{2.28}$$

The general solution of equation 2.26 is obtained by invoking the integrating factor $e^{\tau'_\nu}$

$$I_\nu\left(\tau'_\nu\right) = I_\nu(0)\, e^{-\tau'_\nu} + \int_{\tau''_\nu=0}^{\tau''_\nu=\tau'_\nu} S_\nu\left(\tau''_\nu\right)\, e^{\tau''_\nu-\tau'_\nu}\, d\tau''_\nu. \tag{2.29}$$

The first term on the right-hand side of equation 2.26 is the background intensity $I_\nu(0)$, attenuated by the optical depth τ'_ν to the point in question. The second term represents emission from the medium, i.e. a sum of contributions from all the infinitesimal elements of the medium between $\tau''_\nu = 0$ and $\tau''_\nu = \tau'_\nu$, each one attenuated by the intervening optical depth $\tau'_\nu - \tau''_\nu$.

If the medium is uniform, and hence its emission coefficient and source function are uniform, i.e. $j_\nu\left(\tau'_\nu\right) = j^0_\nu$ and $S_\nu\left(\tau'_\nu\right) = S^0_\nu$, the integral in equation 2.29 is trivial, and we obtain

$$I_\nu\left(\tau'_\nu\right) = I_\nu(0)\, e^{-\tau'_\nu} + S^0_\nu \left\{1 - e^{-\tau'_\nu}\right\}. \tag{2.30}$$

If the background intensity dominates over emission from the medium, then the observed intensity is given by

$$I_\nu\left(\tau_\nu\right) \simeq I_\nu(0)\, e^{-\tau_\nu} \tag{2.31}$$

$$\Rightarrow \tau_\nu \simeq \ln\left[I_\nu(0)/I_\nu\left(\tau_\nu\right)\right]. \tag{2.32}$$

If the background intensity is negligible, and emission from the medium dominates, then the observed intensity is given by

$$I_\nu(\tau_\nu) = S_\nu^0 \left\{ 1 - e^{-\tau_\nu} \right\}. \tag{2.33}$$

Optically thin emission: If the medium is optically thin, $\tau_\nu \ll 1$, the observer receives emission from right through the medium

$$I_\nu(\tau_\nu) \simeq S_\nu^0 \, \tau_\nu = j_\nu^0 \, L. \tag{2.34}$$

Optically thick emission: If the medium is optically thick, $\tau_\nu \gg 1$, almost all the emission received by the observer comes from a thin layer at the front of the medium

$$I_\nu(\tau_\nu) \simeq S_\nu^0. \tag{2.35}$$

In the case where we are studying the continuum emission from dust, the background intensity is usually negligible. If the emitting dust is all at a single temperature, then we can replace S_ν^0 in the last equation with the Planck function $B_\nu(T_{\text{dust}})$

$$
\begin{aligned}
I_\nu &= B_\nu(T_{\text{dust}})[1 - e^{-\tau_\nu}] \\
&= \frac{2h\nu^3}{c^2} \frac{[1 - e^{-\tau_\nu}]}{[e^{(h\nu/kT_{\text{dust}})} - 1]},
\end{aligned}
\tag{2.36}
$$

where T_{dust} is the temperature of the dust (i.e. the temperature characterising the internal vibrations of a dust grain). If we have sufficient measurements of I_ν at different frequencies, we can use this equation to determine the temperature and optical depth of the dust we are observing. This equation is sometimes known as the equation of a greybody.

Observations suggest that in the low-frequency (long-wavelength) limit τ_ν can be approximated by

$$\tau_\nu = \left(\frac{\nu}{\nu_c} \right)^\beta, \tag{2.37}$$

where ν_c is the critical frequency at which the optical depth $\tau_\nu = 1$, β is the dust emissivity index, and typically $1 \le \beta \le 2$. Hence the dust which is optically thick at visible wavelengths becomes optically thin at longer wavelengths. Figure 2.4 shows this effect in the Horsehead Nebula.

2.8 Calculating the dust mass

Study of the emission from dust in the ISM allows us to calculate the mass of the emitting dust as follows. A single, spherical grain of radius a has a monochromatic luminosity L_ν given by

$$L_\nu = 4\pi a^2 \pi \, B_\nu(T_{\text{dust}}) Q_\nu, \tag{2.38}$$

Fig. 2.4. Two pictures of the Horsehead Nebula in Orion. On the left is an optical image, while on the right is a millimetre-wave image (shown in negative, so that bright emission is black). The dust which is absorbing light in the optical image is re-emitting in the millimetre-wave. Furthermore, a bright object in the horse's throat becomes visible in the millimetre-wave, which was obscured in the optical.

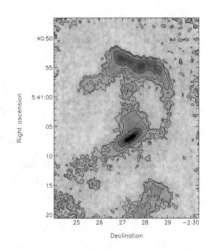

where $4\pi a^2$ is the surface area of the grain, $\pi B_\nu(T_{\text{dust}})$ is the monochromatic flux at frequency ν from a blackbody-like surface with temperature T_{dust}, and Q_ν is the emission efficiency of the grain (i.e. how well it approximates to a blackbody at frequency ν).

If the total mass of dust in the cloud is M_d and the mass of a single dust grain is m_d, then the total number of dust grains in the cloud, $\mathcal{N}_d$, is given by

$$\mathcal{N}_d = \frac{M_d}{m_d} = \frac{3M_d}{4\pi a^3 \rho_d}, \tag{2.39}$$

since $m_d = 4\pi a^3 \rho_d/3$, where ρ_d is the density of the material in a single dust grain.

If the dust emission is optically thin (i.e. the grains at the front of the cloud don't shield the grains at the back) then the flux F_ν received by an observer at distance D is

$$F_\nu = \frac{\mathcal{N}_d L_\nu}{4\pi D^2}, \tag{2.40}$$

Substituting from equations 2.38 and 2.39 into equation 2.40 gives

$$F_\nu = \frac{3M_d B_\nu(T_{\text{dust}})Q_\nu}{4a\rho_d D^2}. \tag{2.41}$$

Rearranging this equation to make M_d the subject, we have

$$M_d = \frac{4a\rho_d F_\nu D^2}{3B_\nu(T_{\text{dust}})Q_\nu}, \tag{2.42}$$

and we can estimate the mass of emitting dust. This is sometimes written as

$$M_d = \frac{F_\nu D^2}{\kappa_d(\nu) B_\nu(T_{\text{dust}})}, \tag{2.43}$$

where $\kappa_d(\nu)$ is known as the dust mass opacity coefficient, given by

$$\kappa_d(\nu) = \frac{3Q_\nu}{4a\rho_d}. \tag{2.44}$$

Long-wavelength emission from dust grains is normally optically thin, and in general the optical depth τ_ν is given by

$$\tau_\nu = N_d \pi a^2 Q_\nu, \tag{2.45}$$

where N_d is the number of dust grains per unit area, also known as the column density. Substituting from equation 2.39 gives

$$\tau_\nu = \frac{3 M_d Q_\nu}{4a\rho_d D^2 \Omega}, \tag{2.46}$$

where Ω is the solid angle subtended at the observer by the emitting dust, and we have used $N_d = \mathcal{N}_d/D^2\Omega$.

Rewriting equation 2.30 for the case of negligible background radiation, and putting $S_\nu = B_\nu(T_{\text{dust}})$, gives

$$I_\nu = B_\nu(T_{\text{dust}})[1 - e^{-\tau_\nu}]. \tag{2.47}$$

Recalling that $F_\nu = I_\nu \Omega$, we have

$$F_\nu = B_\nu(T_{\text{dust}}) \, \Omega \, [1 - e^{-\tau_\nu}], \tag{2.48}$$

and inserting from equation 2.46, gives

$$F_\nu = B_\nu(T_{\text{dust}}) \Omega \left[1 - \exp\left(-\frac{3 M_d Q_\nu}{4a\rho_d D^2 \Omega} \right) \right]. \tag{2.49}$$

Rearranging for M_d gives

$$M_d = -\frac{4a\rho_d D^2 \Omega}{3Q_\nu} \ln\left(1 - \frac{F_\nu}{B_\nu(T_{\text{dust}})\Omega} \right). \tag{2.50}$$

This general equation for all τ_ν reverts to the form of equation 2.42 in the optically thin regime. This is because, from equation 2.48, for small τ_ν, the fraction $[F_\nu / B_\nu(T_{\text{dust}})\Omega] \sim \tau_\nu$, and in the optically thin limit $\tau_\nu \ll 1$.

These very powerful equations illustrate that by measuring the flux from the dust in a molecular cloud we have the potential to calculate the total mass of dust in the cloud. Of course, many assumptions have to be made. In reality the dust grains are not all the same size and are not spherical. However, it can be shown that the effects of these considerations can be minimised if one observes at a frequency such that the wavelength $\lambda \gg a$, and if a mean grain size $\bar{a}$ is used.

Mean grain sizes can be estimated in various ways, such as from their refractory and polarisation properties, as can the bulk material properties of the dust, such as ρ_d. Measurements at many different frequencies can

Fig. 2.5. Bound energy levels.

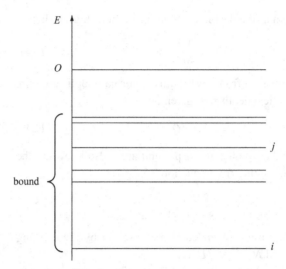

help tie down T_{dust} and τ_ν, and so an estimate of the mass of dust in a cloud can be made.

2.9 Line radiation – studying the gas

Consider a species X (atom or ion or molecule) which has two energy levels, labelled i and j, having energies E_i and E_j, where E_j is the higher level, and $h\nu_0 = E_j - E_i$ is the energy difference between the levels – see Figure 2.5. For simplicity, we assume that these are the only energy levels, but this does not affect the generality of the results we obtain. Now suppose that there is a gas of this species together with free electrons having number density n_e and continuum radiation having energy density u_{ν_0}.

2.9.1 Population transfer

Transfer of population directly between these energy levels occurs by a variety of processes:

Spontaneous radiative de-excitation (also called spontaneous emission) transfers population downward at a rate per unit volume equal to $(n_{X;j} A_{ji})$, where $n_{X;j}$ is the number density of particles of species X in the upper level j, and A_{ji} is the Einstein A-coefficient.

Induced radiative de-excitation (also called induced or stimulated emission) transfers population downward at a rate per unit volume equal to $(n_{X;j} B_{ji} u_{\nu_0})$, where B_{ji} is one of the Einstein B-coefficients.

Collisional de-excitation transfers population downward at a rate per unit volume equal to $(n_{X;j} C_{ji}(T) n_\mathrm{e})$, where $C_{ji}(T)$ is the collisional de-excitation coefficient due to electrons, and n_e is the number density of free electrons. We have assumed that electrons are the main agents of collisional de-excitation, and this is usually the case, unless there are very few free electrons. If the density of free electrons is low, then the job must be done by hydrogen atoms, helium atoms, or hydrogen molecules.

Note that spontaneous de-excitation does not require an external agent. It happens at the same rate irrespective of environment. In contrast, the rate of induced radiative de-excitation depends on the ambient intensity of the radiation field at frequency ν_0, and the rate of collisional de-excitation depends on the density and temperature of the colliding particles (e.g. electrons).

Radiative excitation (also called radiative absorption) transfers population upwards at a rate per unit volume given by

$$n_{X;i} \, B_{ij} \, u_{\nu_0} = n_{X;i} \, \sigma_0 \, \frac{c \, u_{\nu_0}}{h \nu_0}, \tag{2.51}$$

where B_{ij} is the second Einstein B-coefficient, and σ_0 is the integrated absorption cross-section.

Collisional excitation transfers population upwards at a rate per unit volume equal to $(n_{X;i} C_{ij}(T) n_e)$, where $C_{ij}(T)$ is the collisional excitation coefficient, and again we are assuming that collisions with electrons are the dominant collisional excitation mode.

2.9.2 Population distributions

In thermodynamic equilibrium (TE), the gas particles have a velocity distribution that is Maxwellian in form, i.e. the number of particles of species X in unit volume having velocity components in the range $(u_x, u_x + du_x; \, u_y, u_y + du_y; \, u_z, u_z + du_z)$ is given by

$$n_{X;u_x,u_y,u_z} \, du_x \, du_y \, du_z = n_X \left[\frac{m_X}{2\pi kT} \right]^{3/2}$$
$$\times \exp \left[-\frac{m_X \left(u_x^2 + u_y^2 + u_z^2 \right)}{2kT} \right] du_x \, du_y \, du_z, \tag{2.52}$$

where n_X is the total number density of particles of species X, and m_X is the mass of a single particle of species X. Similarly, the number of particles of species X in unit volume having speed in the range $(u, u + du)$ is

$$n_{X;u} \, du = n_X \left[\frac{m_X}{2\pi kT} \right]^{3/2} \exp \left[-\frac{m_X u^2}{2kT} \right] 4\pi u^2 du. \tag{2.53}$$

If the species X has two internal energy levels i and j (see Figure 2.5), having energies E_i and E_j, and statistical weights g_i and g_j, the relative populations of the two levels in TE are given by the Boltzmann equation

$$\frac{n_{X;j}}{n_{X;i}} = \frac{g_j \exp[-E_j/kT]}{g_i \exp[-E_i/kT]} = \frac{g_j}{g_i} \exp\left[\frac{-\left(E_j - E_i\right)}{kT}\right]$$

$$= \frac{g_j}{g_i} \exp\left[-\frac{h\nu_0}{kT}\right], \tag{2.54}$$

where we have simplified the expression using the frequency of the transition.

It follows that the fraction of particles of species X in level j is given by

$$\frac{n_{X;j}}{n_X} = \frac{g_j \exp[-E_j/kT]}{Z_X(T)}, \tag{2.55}$$

where $Z_X(T)$ is called the partition function of species X (Z stands for *Zustandsumme*, which is German for 'sum of states'), and is given by

$$Z_X(T) = \sum_{\text{all levels } i} \{g_i \exp[-E_i/kT]\}. \tag{2.56}$$

2.9.3 The Einstein relations between coefficients

We now consider only a single species. In thermodynamic equilibrium (TE) for our two-level atom we have

$$\frac{n_{X;i}}{n_{X;j}} = \frac{g_i \exp\left[-E_i/kT\right]}{g_j \exp\left[-E_j/kT\right]} = \frac{g_i}{g_j} \exp\left[\frac{h\nu_0}{kT}\right], \tag{2.57}$$

from the Boltzmann equation, and

$$u_{\nu_0} = \frac{4\pi B_{\nu_0}(T)}{c} = \frac{8\pi h\nu_0^3}{c^3} \left\{\exp\left[\frac{h\nu_0}{kT}\right] - 1\right\}^{-1}, \tag{2.58}$$

but statistical equilibrium requires an exact balance between the rate at which population is transferred upwards from level i to level j and the rate at which population is transferred downwards from level j to level i

$$n_{X;j} \left\{A_{ji} + B_{ji} \frac{4\pi B_{\nu_0}(T)}{c} + C_{ji}(T)n_e\right\}$$

$$= n_{X;i} \left\{B_{ij} \frac{4\pi B_{\nu_0}(T)}{c} + C_{ij}(T)n_e\right\}. \tag{2.59}$$

Moreover, this balance must hold at all temperatures, and all densities, i.e. for all T and all n_e.

Since the coefficients A_{ji}, B_{ji}, B_{ij}, $C_{ji}(T)$ and $C_{ij}(T)$ are intrinsic properties of the particles, i.e. properties which remain the same, whatever the surrounding density, we can argue as follows.

First, suppose that, with the temperature held fixed (but at an arbitrary value), we vary the density. The collisional terms in equation 2.59 involve the square of the density, and so they must balance independently, i.e. irrespective of the radiative terms. This is because we can increase the density, and thereby make these collisional terms arbitrarily large compared with the other terms, but the balance still has to hold. Therefore we must have

$$C_{ji}(T) = \frac{n_{X;i}}{n_{X;j}} \, C_{ij}(T) = \frac{g_i}{g_j} \exp\left[\frac{h\nu_0}{kT}\right] C_{ij}(T), \qquad (2.60)$$

at all temperatures T (where we have substituted for $n_{X;i}/n_{X;j}$ from equation 2.57).

Second, if the collisional terms in equation 2.59 balance independently, the radiative terms in equation 2.59 must also balance independently. After all, we can make the collisional terms arbitrarily small by decreasing the density. Therefore we have

$$A_{ji} + \frac{8\pi h}{\lambda_0^3} \left\{\exp\left[\frac{h\nu_0}{kT}\right] - 1\right\}^{-1} B_{ji}$$

$$= \frac{n_{X;i}}{n_{X;j}} \frac{8\pi h}{\lambda_0^3} \left\{\exp\left[\frac{h\nu_0}{kT}\right] - 1\right\}^{-1} B_{ij}$$

$$= \frac{g_i}{g_j} \exp\left[\frac{h\nu_0}{kT}\right] \frac{8\pi h}{\lambda_0^3} \left\{\exp\left[\frac{h\nu_0}{kT}\right] - 1\right\}^{-1} B_{ij}, \qquad (2.61)$$

where again we have substituted for $n_{X;i}/n_{X;j}$ from equation 2.57. Then multiplying this equation throughout by the term in curly brackets, we obtain

$$A_{ji} \left\{\exp\left[\frac{h\nu_0}{kT}\right] - 1\right\} + \frac{8\pi h}{\lambda_0^3} B_{ji} = \frac{g_i}{g_j} \exp\left[\frac{h\nu_0}{kT}\right] \frac{8\pi h}{\lambda_0^3} B_{ij}, \qquad (2.62)$$

again at all temperatures T.

Third, the terms in equation 2.62 which involve the temperature must balance independently of the others. This is because the terms involving the temperature can be made arbitrarily large by decreasing the temperature. Therefore we have

$$B_{ij} = \frac{\lambda_0^3}{8\pi h} \frac{g_j}{g_i} A_{ji}. \qquad (2.63)$$

Fourthly (and finally), the terms in equation 2.62 not involving the temperature must also balance independently, giving

$$B_{ji} = \frac{\lambda_0^3}{8\pi h} A_{ji} = \frac{g_i}{g_j} B_{ij}. \tag{2.64}$$

Equations 2.60, 2.63 and 2.64 are known as Einstein's relations. They are very useful, because they mean that quantum physicists only need to compute one collisional coefficient – say $C_{ij}(T)$ – and one radiative coefficient – say A_{ji}. The other three coefficients are then obtained trivially using Einstein's relations.

2.9.4 Emission and absorption coefficients

The absorbing power of a transition can be measured by its integrated cross-section, σ_0, which is related to B_{ij} (see equation 2.51) by

$$\sigma_0 = \frac{h B_{ij}}{\lambda_0}. \tag{2.65}$$

The integrated emission coefficient for line radiation (due to the transition discussed above) is given by

$$j = \frac{n_{X;j} A_{ji} h\nu_0}{4\pi}, \tag{2.66}$$

where $n_{X;j}$ is the number of particles of species X in level j, in unit volume. A_{ji} is the rate at which these particles de-excite radiatively and spontaneously to level i, and so $n_{X;j} A_{ji}$ is the number of line photons emitted spontaneously, from unit volume, in unit time. The factor $h\nu_0$ converts this into the amount of *radiant energy* emitted in the line, from unit volume, in unit time (see Section 2.1). The factor $1/4\pi$ converts this into the amount of radiant energy emitted in the line, from unit volume, in unit time, *into unit solid angle* (on the assumption that the emission is isotropic). This is, by definition, the integrated emission coefficient.

The monochromatic emission coefficient is then given by

$$j_\nu = j \times \phi (\nu - \nu_0), \tag{2.67}$$

where $\phi(\nu - \nu_0)$ is called the profile function, and measures how the emission is distributed about the central frequency. Since

$$j = \int_{\text{line}} j_\nu \, d\nu = \int_{\text{line}} j \, \phi(\nu - \nu_0) \, d\nu = j \int_{\text{line}} \phi(\nu - \nu_0) \, d\nu, \tag{2.68}$$

the profile function must be normalised,

$$\int_{\nu=0}^{\nu=\infty} \phi(\nu - \nu_0) \, d\nu = 1, \tag{2.69}$$

and so $\phi(\nu - \nu_0)$ has the dimensions of 'one over frequency', i.e. time.

The opacity coefficient is given by

$$\kappa_\nu = n_{X;i}\, \sigma_0\, \phi\, (\nu - \nu_0) \left\{ 1 - \frac{n_{X;j} B_{ji}}{n_{X;i} B_{ij}} \right\}$$

$$= n_{X;i}\, \sigma_0\, \phi\, (\nu - \nu_0) \left\{ 1 - \frac{n_{X;j} g_i}{n_{X;i} g_j} \right\}. \tag{2.70}$$

Here, $n_{X;i}\sigma_0\phi(\nu - \nu_0)$, gives the *true* absorption. The terms inside the curly brackets represent a correction factor to account for stimulated emission, which acts like negative absorption. To obtain the final expression, we have used equation 2.64 to replace B_{ji}/B_{ij} with g_i/g_j.

In thermodynamic equilibrium, $n_{X;j}/n_{X;i}$ is given by the Boltzmann equation (equation 2.57), and so the correction factor for stimulated emission reduces to

$$\left\{ 1 - \frac{n_{X;j} g_i}{n_{X;i} g_j} \right\} \rightarrow \left\{ 1 - \exp\left[-\frac{h\nu_0}{kT} \right] \right\}. \tag{2.71}$$

If $h\nu_0 \ll kT$, then this approximates to

$$\left\{ 1 - \exp\left[-\frac{h\nu_0}{kT} \right] \right\} \simeq \left\{ \frac{h\nu_0}{kT} \right\} \ll 1, \tag{2.72}$$

and so under this circumstance the effective opacity coefficient is much smaller than the true absorption coefficient; in other words, almost every true absorption is balanced by a stimulated emission.

Recommended further reading

We recommend the following texts to the student for further reading on the topics presented in this chapter.

Emerson, D. (1996). *Interpreting Astronomical Spectra*. New York: Wiley

Krugel, E. (2003). *The Physics of Interstellar Dust*. Bristol: Institute of Physics Press.

Shu, F. H. (1991). *The Physics of Astrophysics – Radiation*, vol. 1. Mill Valley: University Science Books.

Whittet, D. (1992). *Dust in the Galactic Environment*. Bristol: Institute of Physics Press.

Chapter 3
The ISM – the beginnings of star formation

3.1 Introduction

In this chapter we take a more detailed look at the interstellar medium (ISM). We consider first the most abundant element in the Universe, hydrogen. We discuss the atomic hydrogen transition which occurs at 21 cm. We look at the 21-cm line in both absorption and emission. We then go on to consider the molecular gas and, in particular, the most abundant gas-phase molecule after hydrogen, carbon monoxide (CO). We also look at the use of absorption lines in the study of the ISM. In this context we consider some features of spectral lines, such as their equivalent widths, and we describe the curve of growth of a spectral line. In the next chapter we will go on to study the denser parts of the ISM, known as molecular clouds.

3.2 The 21-cm line of atomic hydrogen

The most abundant element in the Universe is hydrogen. We here discuss the main signature of cool atomic hydrogen, 21-cm line radiation. Figure 3.1 shows 21-cm images of some nearby galaxies, illustrating how the 21-cm radiation traces the atomic gas in the interstellar medium of these galaxies.

The electronic ground level of atomic hydrogen has quantum numbers: $n = 1, l = 0, m_l = 0$ and $m_s = \pm 1/2$. To first order the energy of this level is determined solely by the principal quantum number n

$$E_n = -\frac{2\pi^2 m_e e^4}{h^2 n^2} \quad \xrightarrow{n=1} \quad E_1 = -\frac{2\pi^2 m_e e^4}{h^2}. \tag{3.1}$$

Fig. 3.1. Images of other galaxies taken in the 21-cm line of atomic hydrogen (HI). The images are tracing the atomic gas in the interstellar medium of these galaxies.

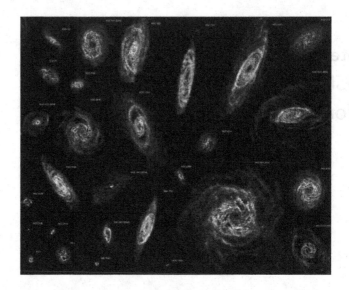

This level is fourfold degenerate, because not only does the electron have spin ($s = 1/2$, giving $m_s = \pm 1/2$), but the proton also has spin ($i = 1/2$, giving $m_i = \pm 1/2$).

3.2.1 21-cm energy levels

The fourfold degeneracy of the ground level is partially raised by hyperfine splitting due to the interaction between the magnetic moments of the electron and the proton. We now explain how the hyperfine splitting works.

In the true ground state, which we denote with a subscript 0, the spins and magnetic moments are antiparallel, so the total angular momentum f_0 is zero and the statistical weight g_0 is one

$$f_0 = |s - i| = 0, \quad \Longrightarrow \quad m_f = 0, \quad g_0 = 1. \tag{3.2}$$

In the excited level, which we denote with a subscript 1, the spins and magnetic moments are parallel, so the total angular momentum f_1 is one and the statistical weight g_1 is three

$$f_1 = |s + i| = 1, \quad \Longrightarrow \quad m_f = 0, \pm 1, \quad g_1 = 3. \tag{3.3}$$

We note parenthetically that one cannot think of the magnetic moments associated with the electron and proton as current loops due to extended rotating distributions of charge. If this were the case, the magnetic moment of the electron would be antiparallel to its spin, and the magnetic moment of the proton would be parallel to its spin. In fact, in both cases the magnetic moment is parallel to the spin.

The interaction energy due to the magnetic moments is very small

$$\Delta E \equiv E_1 - E_0 \simeq 10^{-24}\,\text{J}$$

$$\simeq k\,[0.07\,\text{K}] \simeq h\,[1.4\,\text{GHz}] \simeq \frac{hc}{[21\,\text{cm}]}, \tag{3.4}$$

where the last three expressions give the energy in terms of (i) temperature (using the Boltzmann constant k), (ii) frequency and (iii) wavelength (21 cm $= 0.21$ m).

The spontaneous radiative de-excitation coefficient (Einstein A-coefficient) is

$$A_{10} \simeq 3 \times 10^{-15}\,\text{s}^{-1} \simeq \left[10^7\,\text{years}\right]^{-1}. \tag{3.5}$$

In other words, a typical hydrogen atom spends on average $\sim 10^7$ years in the upper level j before it de-excites spontaneously, with the emission of a 21-cm photon. One might conclude that this radiation would be undetectable. However, the weak emission rate per hydrogen atom is compensated by the enormous amount of atomic hydrogen in interstellar space.

3.2.2 21-cm level populations

For any transition, we define the excitation temperature, T_{ex}, as the temperature at which the Boltzmann distribution of population between the two levels (i.e. the thermodynamic equilibrium distribution) equals the actual distribution. In this case we have

$$\frac{n_1}{n_0} = \frac{g_1}{g_0}\exp\left[-\frac{\Delta E}{kT_{ex}}\right] = 3\exp\left[-\frac{0.07\,\text{K}}{T_{ex}}\right]$$

$$\simeq 3 - \frac{0.21\,\text{K}}{T_{ex}} \sim 3. \tag{3.6}$$

In the equation above, we have introduced two stages of approximation, which are valid as long as $T_{ex} \gg 0.07$ K. Typically $T_{ex} \stackrel{>}{\sim} 80$ K, so this is a safe assumption. The first stage, denoted by $\simeq$, retains only terms of zeroth and first order in $(1/T_{ex})$. The second stage, denoted by $\sim$, retains only zeroth-order terms.

Since collisions dominate the transfer of population between the two levels, we have $T_{ex} \simeq T$. In other words, the frequency and energetics of collisions reflect the velocity distribution of the particles, and this approximates closely to its thermodynamic equilibrium form, which is a Maxwellian distribution at temperature T. Consequently the distribution of population also approximates to its thermodynamic equilibrium form, which is a Boltzmann distribution at the same temperature T.

After the two levels involved in the 21-cm transition, the next excited level is the level with principal quantum number $n = 2$ and energy E_2, where

$$E_2 - E_1 \simeq 1.6 \times 10^{-18} \text{ J} \simeq k \left[1.2 \times 10^5 \text{ K} \right]. \tag{3.7}$$

Therefore, for $T \overset{<}{\sim} 10^3$ K, we can neglect the population in this and higher levels, and put

$$n_0 + n_1 = n_{\text{HI}}, \tag{3.8}$$

where n_0 is the number of particles per unit volume in the true ground state ($f = 0$), n_1 is the number of particles per unit volume in the excited level ($f = 1$), and n_{HI} is the total number of neutral hydrogen (HI) atoms per unit volume.

The partition function is therefore given by

$$Z_X \simeq g_0 + g_1 \exp \left[\frac{-\Delta E}{kT} \right] = 1 + 3 \exp \left[\frac{-0.07 \text{ K}}{T} \right]$$

$$\simeq 4 - \frac{0.21 \text{ K}}{T} \sim 4, \tag{3.9}$$

where we have again employed the two stages of approximation defined earlier.

It follows that the level populations are given by

$$\frac{n_0}{n_{\text{HI}}} \simeq \frac{g_0}{Z_X} \simeq \frac{1}{\left[4 - \frac{0.21 \text{ K}}{T} \right]} \simeq 0.25 + \frac{0.01331 \text{ K}}{T} \sim 0.25,$$

and

$$\frac{n_1}{n_{\text{HI}}} \simeq \frac{g_1 \exp \left[\frac{-\Delta E}{kT} \right]}{Z_X} \simeq \frac{\left[3 - \frac{0.21 \text{ K}}{T} \right]}{\left[4 - \frac{0.21 \text{ K}}{T} \right]}$$

$$\simeq 0.75 - \frac{0.01331 \text{ K}}{T} \sim 0.75. \tag{3.10}$$

In other words, slightly more than a quarter of the population is in the true ground state ($f = 0$), and slightly less than three quarters is in the excited level ($f = 1$).

3.2.3 Radiative transfer in the 21-cm line

The monochromatic volume opacity coefficient for 21-cm line radiation can be derived from equations 2.70 and 3.6 to be

$$\kappa_\nu = n_0 \sigma_0 \phi(\nu - \nu_0) \left\{ 1 - \exp \left[\frac{-0.07 \text{ K}}{T} \right] \right\}. \tag{3.11}$$

With the first stage of approximation, this becomes

$$\kappa_\nu = n_0 \sigma_0 \phi(\nu - \nu_0) \left\{ \frac{0.07 \text{ K}}{T} \right\}, \tag{3.12}$$

so we see that the correction for stimulated emission, the term in braces, is usually rather significant. At $T \simeq 100$ K, a typical temperature for atomic hydrogen in the interstellar medium, there are about 1400 stimulated emissions for every 1401 true absorptions.

Putting

$$n_0 \sim \frac{n_{HI}}{4}, \tag{3.13}$$

from equation 3.10, and

$$\sigma_0 = \frac{g_1 c^2 A_{10}}{8 \pi g_0 v^2} \simeq 1.6 \times 10^{-17} \text{ m}^2 \text{ s}^{-1}, \tag{3.14}$$

from equations 2.64, 2.65 and 3.5, equation 3.12 becomes

$$\kappa_v \simeq C \frac{n_{HI}}{T} \phi(v - v_0), \tag{3.15}$$

with

$$C = \frac{g_1 c^2 h A_{10}}{32 \pi g_0 k v_0} = 3 \times 10^{-19} \text{ m}^2 \text{ s}^{-1} \text{ K}. \tag{3.16}$$

For a line of sight dominated by a single uniform cloud (in particular, a cloud with uniform temperature), we obtain

$$\tau_v \simeq \int_{l=0}^{l=\infty} \kappa_v(l) \, dl \simeq C \frac{N_{HI}}{T} \phi(v - v_0), \tag{3.17}$$

where N_{HI} is the column density of hydrogen atoms through the cloud, defined as the number of atoms in a column of unit area that extends along the line of sight through the entire cloud. This is often a more useful concept than the volume density because it is more directly related to what is actually observed, and does not rely on assumptions about the distribution of the emitting gas along the line of sight.

Equation 3.17 shows that cool clouds absorb more effectively than warm ones. This is because the imbalance between stimulated emission and true absorption increases with decreasing temperature (see equation 3.12). Conversely, as the temperature is increased to sufficiently large values,

$$\frac{n_1}{n_0} \longrightarrow \frac{g_1}{g_0}, \tag{3.18}$$

and the net absorption (true absorption minus stimulated emission) tends to zero.

From equations 2.66 and 2.67, the monochromatic volume emission coefficient, giving the rate of spontaneous emission of energy from unit volume into unit solid angle, is

$$j\phi(v - v_0) = \frac{n_1 A_{10} h v_0 \phi(v - v_0)}{4\pi}. \tag{3.19}$$

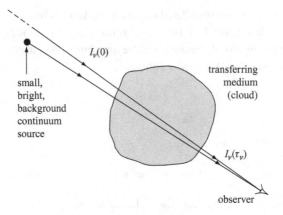

If we combine equations 3.19 and 3.11, we find, after a little algebra,

$$S_\nu = \frac{j_\nu}{\kappa_\nu} = ... \longrightarrow ..... B_\nu(T) \simeq \frac{2kT\nu_0^2}{c^2}. \tag{3.20}$$

Here $B_\nu(T)$ is the Planck function for the intensity of blackbody radiation at the excitation temperature T (remember that the thermodynamic temperature is equal to the excitation temperature, since collisions dominate the population transfer). The last expression gives the Rayleigh–Jeans limit, which is valid when $h\nu \ll kT$. We have substituted $\nu \to \nu_0$ because the range of frequency across a line is very small.

It is not coincidental that the source function approximates to the Planck function at temperature T. It happens because the levels have a population corresponding to thermodynamic equilibrium at temperature T (see Section 3.2.2). Since the source function is determined entirely by the level populations, it must adopt the value corresponding to thermodynamic equilibrium at temperature T – i.e. the Planck function $B_\nu(T)$.

3.2.4 The 21-cm line in absorption

Figure 3.2 illustrates the ideal situation in which there is a small but resolved bright background continuum source behind a single cool cloud (denoted on the figure as the transferring medium). We make observations on the line of sight to the resolved background source, and on a line of sight which misses the background source. We assume that the cloud is uniform, in particular the volume emission coefficient, volume opacity coefficient, and source function are independent of position

$$j_\nu(l) = j_\nu^0, \tag{3.21}$$

$$\kappa_\nu(l) = \kappa_\nu^0, \tag{3.22}$$

$$S_\nu(l) = S_\nu^0. \tag{3.23}$$

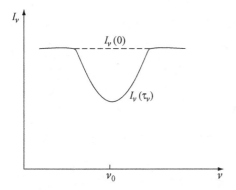

Fig. 3.3. 21-cm absorption
line in the spectrum of a
bright background continuum
source.

From equation 2.30, the general solution of the equation of radiation
transport through a uniform transporting medium is

$$I_\nu(\tau_\nu) = I_\nu(0)\, e^{-\tau_\nu} + S_\nu^0 \left[1 - e^{-\tau_\nu} \right]. \qquad (3.24)$$

Here $I_\nu(0)$ is the background intensity incident on the far side of
the medium, and $I_\nu(\tau_\nu)$ is the intensity of the radiation emerging on
the observer's side of the medium. The first term on the right-hand
side represents the background radiation attenuated by passing though
the medium (see Section 2.7). The second term represents emission from
the medium itself. This emission is a sum of contributions from each
infinitesimal layer of the medium, with each such layer being attenuated
by the amount of medium which lies in front of it.

On the line of sight to the bright background continuum source,
the emission from the medium can be neglected. Therefore we see an
absorption line (as illustrated in Figure 3.3) with

$$I_\nu(\tau_\nu) = I_\nu(0)\, e^{-\tau_\nu}. \qquad (3.25)$$

From equation 3.17, we deduce that

$$\tau_\nu \simeq C\, \frac{N_{\mathrm{HI}}}{T}\, \phi(\nu - \nu_0) \simeq \ln\left[\frac{I_\nu(0)}{I_\nu(\tau_\nu)} \right]. \qquad (3.26)$$

By integrating over frequency, we obtain

$$C\, \frac{N_{\mathrm{HI}}}{T} \int_{\mathrm{line}} \phi(\nu - \nu_0)\, d\nu = C\, \frac{N_{\mathrm{HI}}}{T} = \int_{\mathrm{line}} \ln\left[\frac{I_\nu(0)}{I_\nu(\tau_\nu)} \right] d\nu$$

$$\implies \quad \frac{N_{\mathrm{HI}}}{T} \simeq \frac{1}{C} \int_{\mathrm{line}} \ln\left[\frac{I_\nu(0)}{I_\nu(\tau_\nu)} \right] d\nu. \qquad (3.27)$$

Then, if we measure the observed intensity $I_\nu(\tau_\nu)$, and estimate
the background intensity $I_\nu(0)$ (by interpolating across the line – see
Figure 3.3), we can perform the integral on the right-hand side of equation 3.27, and hence evaluate the combination N_{HI}/T on the line of sight

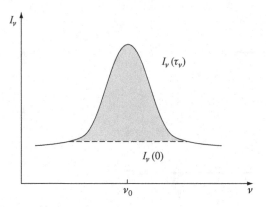

Fig. 3.4. Optically thin 21-cm emission line.

to the background source. However, if the line is very optically thick, so that the centre is completely wiped out, $I_\nu(\tau_\nu)$ will be poorly known, and so the estimate of $N_{\rm HI}/T$ will be rather uncertain.

3.2.5 The 21-cm line in emission

If we now point our telescope just to the side of the bright background continuum source, the background intensity is much smaller, and so we see the line in emission (see Figure 3.4). From equation 3.24 we have

$$I_\nu(\tau_\nu) = I_\nu(0)\,{\rm e}^{-\tau_\nu} + S_\nu^0\left[1 - {\rm e}^{-\tau_\nu}\right]. \qquad (3.28)$$

If the line is optically thin, $\tau_\nu \ll 1$, the observed intensity $I_\nu(\tau_\nu)$ will be given by

$$I_\nu(\tau_\nu) - I_\nu(0) \simeq S_\nu^0\,\tau_\nu = \frac{j_\nu^0}{\kappa_\nu^0}\,\kappa_\nu^0 L = j_\nu^0 L$$

$$= \frac{3N_{\rm HI}A_{10}\Delta E}{16\pi}\,\phi(\nu - \nu_0). \qquad (3.29)$$

In other words, we will see emission from throughout the emitting medium; there will be no significant shielding of the emission from the HI atoms at the back of the cloud by those at the front. Equation 3.29 shows that the line shape will echo the profile function $\phi(\nu - \nu_0)$ – see Figure 3.4.

If we now integrate over the line, we obtain the integrated intensity of the line

$$I_{\rm line} \equiv \int_{\rm line}\left[I_\nu(\tau_\nu) - I_\nu(0)\right]d\nu$$

$$= \frac{3N_{\rm HI}A_{10}\Delta E}{16\pi}\int_{\rm line}\phi(\nu - \nu_0)\,d\nu$$

$$= \frac{3N_{\rm HI}A_{10}\Delta E}{16\pi}. \qquad (3.30)$$

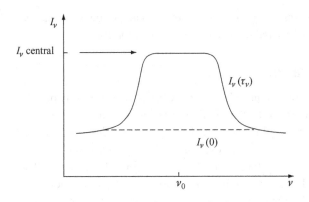

Fig. 3.5. Optically thick 21-cm emission line.

Hence we can evaluate the column density of atomic hydrogen from the equation

$$N_{HI} \simeq \frac{16\pi}{3A_{10}\Delta E} I_{line} \simeq \left[5 \times 10^{39} \text{ s ster J}^{-1}\right] I_{line}. \tag{3.31}$$

Because this is a vital aspect of the discussion, we reiterate that the integrated intensity is proportional to the column density of emitting particles only if the line is optically thin (no shielding).

As long as we are happy to assume that the column of atomic hydrogen on the line of sight to the bright background continuum source, and the column on the nearby line of sight which misses this source, are similar (as we have drawn them in Figure 3.2), then we can combine the results from equations 3.27 and 3.31 to obtain both the column density of atomic hydrogen N_{HI} and the gas kinetic temperature T.

If the line is optically thick, $\tau_\nu \gg 1$, over a substantial range of frequencies at the line centre, equation 3.24 can be approximated by

$$I_\nu(\tau_\nu) \simeq S_\nu(0) \simeq B_\nu(T) \simeq \frac{2kT\nu^2}{c^2}. \tag{3.32}$$

In this case, the intensity at the centre of the line is approximately independent of frequency (see Figure 3.5). In the centre of the line we are only seeing emission from the front layers of the cloud; the layers behind are shielded. From equation 3.32, we can write

$$T \simeq \frac{c^2 I_{\nu_0}}{2k\nu_0^2} \simeq \left[1.6 \times 10^{21} \text{K ster m}^2\text{J}^{-1}\right] I_{\nu_0}, \tag{3.33}$$

where I_{ν_0} is the intensity at the centre of the line.

Again, as long as we are happy to assume that the columns of atomic hydrogen on the two lines of sight are similar, then we can combine the results from equations 3.27 and 3.33 to obtain both the column density of atomic hydrogen N_{HI} and the gas kinetic temperature T. However, as we have already noted, in this case the combination N_{HI}/T will be rather uncertain, and so N_{HI} will also be uncertain. Thus we require several independent mass estimates to agree before we can be confident that we have accurately measured the mass of a region of the ISM.

3.3 Molecular gas

Stars form in regions of the ISM where the gas is predominantly molecular. These regions are known as molecular clouds, and they will be discussed in more detail in the next chapter. Here we describe how the physical parameters of such regions are derived from observations.

The angular momentum of a molecule can only have certain discrete values, and therefore its rotational energy is quantised. These values, $E(J)$, are described by the angular momentum quantum number J: $E(J) = \hbar^2 J(J+1)/2\mathcal{I}$, where $\mathcal{I}$ is the moment of inertia of the molecule. The usual method of observing molecular gas in the ISM is by way of the rotational transitions of the molecules between different J levels. The most common transitions are the electric dipole transitions, for which $\Delta J = \pm 1$.

3.3.1 The problems of detecting H_2

Since hydrogen is the most abundant element in the Universe, the most abundant molecule is H_2. However, H_2 is a homonuclear diatomic molecule (a symmetric dumb-bell) so it has no permanent electric dipole moment. Therefore its electric dipole transitions are forbidden, and we must look to the next set of allowed transitions.

The selection rules for electric quadrupole transitions in H_2 are $\Delta J = 0, \pm 2$. Consequently there are two forms of H_2: para-H_2, which can only occupy the rotational states with even quantum numbers $J = 0, 2, 4, 6$, etc.; and ortho-H_2, which can only occupy the rotational states with odd quantum numbers $J = 1, 3, 5$, etc. Conversion of para-H_2 into ortho-H_2, or vice versa, occurs only very slowly at low temperatures (in the absence of a catalyst).

The first rotationally excited state of para-H_2 is at

$$\Delta E \equiv E(J = 2) - E(J = 0) = \frac{3h^2}{4\pi^2 \mathcal{I}_{H_2}}$$

$$\simeq 7.5 \times 10^{-21} \text{J} \simeq k \, [540 \, \text{K}], \tag{3.34}$$

where

$$\mathcal{I}_{H_2} \simeq 5 \times 10^{-48} \text{ kg m}^2 \tag{3.35}$$

is the moment of inertia of an H_2 molecule. At the typical temperatures observed in regions of molecular gas in the ISM ($T \sim 10$–40 K) H_2 is not significantly excited. Therefore it does not in general emit much radiation.

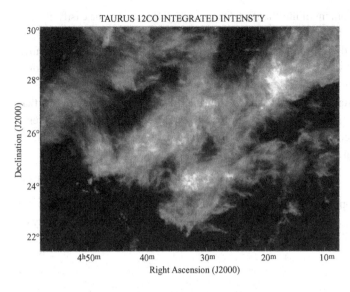

TAURUS 12CO INTEGRATED INTENSTY

Fig. 3.6. Carbon monoxide (CO) image of the molecular cloud in the constellation Taurus. CO is used as a tracer of molecular gas in general. A great deal of structure can be seen in the cloud.

H$_2$ can be detected via its UV *absorption* lines (the same lines that are involved in its destruction), but only on the lines of sight to a few suitably positioned bright, hot, background stars. These lines of sight are too few and far between to make this an effective means of conducting either a global search for, or a survey of, H$_2$.

3.3.2 Using CO to trace H$_2$

Since H$_2$ is so hard to observe, carbon monoxide (CO) is used as a tracer of molecular gas. The advantages of CO are that it is an abundant and relatively stable molecule, and therefore it has a high abundance compared with most other molecules. In addition, it has a small dipole moment, so its transitions are electric-dipole allowed. Finally, because it is quite a massive molecule, it has a large moment of inertia, and so its rotational energy levels are closely spaced and easily excited at the low temperatures encountered in molecular cloud regions of the ISM. Figure 3.6 shows an image made of the molecular cloud in the constellation Taurus, using the CO molecule as a tracer.

The first rotationally excited level of CO has energy

$$\Delta E \equiv E(J = 1) - E(J = 0) = \frac{h^2}{4\pi^2 \mathcal{I}_{\mathrm{CO}}}$$

$$\simeq 7.6 \times 10^{-23}\mathrm{J} = k\,[5.5\,\mathrm{K}]$$

$$\simeq h\,[10^{11}\,\mathrm{Hz}] \simeq hc\,[0.0026\,\mathrm{m}]^{-1} \tag{3.36}$$

where

$$\mathcal{I}_{\mathrm{CO}} \simeq 1.5 \times 10^{-46}\ \mathrm{kg\ m^2} \tag{3.37}$$

is the moment of inertia of a CO molecule. This line can be easily excited at low temperatures ($T \sim 10$–40 K).

Furthermore, at volume number densities $n \gtrsim 10^9$ H$_2$ molecules per m^3, such as are found in molecular clouds, collisions tend to dominate the transfer of population between the levels $J = 1$ and $J = 0$. This establishes a Boltzmann-like distribution of population, so that the excitation temperature approximates closely to the gas kinetic temperature, $T_{ex} \simeq T$, and the source function approximates to the Planck function at temperature T

$$S_\nu \simeq B_\nu(T). \tag{3.38}$$

Suppose that we observe line emission at $\lambda \simeq 2.6$ mm, which we attribute to the $J = 1 \rightarrow 0$ transition of CO. The integrated intensity of the line is related to the monochromatic intensity by

$$I = \int_{\text{line}} I_\nu \, d\nu. \tag{3.39}$$

However, observers find it convenient to replace the intensity I_ν with the brightness temperature T_B, which is the temperature at which a blackbody radiation field has the same intensity. In other words $B_\nu(T_B) = I_\nu$. For the relatively low frequencies of millimetre lines like the CO $J = 1 \rightarrow 0$ line, the Planck function approximates to $B_\nu(T_B) \simeq 2kT_B\nu^2/c^2$, and so

$$T_B \simeq \frac{c^2 I_\nu}{2k\nu^2}, \tag{3.40}$$

although the results we shall present below do not depend on this approximation. Likewise, the frequency ν is usually replaced with the corresponding radial line-of-sight velocity v, which is related to the frequency by the Doppler shift, $v/c = (\nu_0 - \nu)/\nu_0$, where ν_0 is the rest frequency of the line.

Making the above replacements, the integrated intensity of the CO ($J = 1 \rightarrow 0$) line becomes

$$I_{CO} = \int_{\text{line}} T_B(v) \, dv, \tag{3.41}$$

which is normally quoted in units of K km s^{-1}.

For the ^{12}C^{16}O ($J = 1 \rightarrow 0$) line, the line centre is usually optically thick.[†] Therefore the intensity approximates to the source function, which is $S_\nu \simeq B_\nu(T)$ (see equation 3.38), giving $T_B \simeq T$. Hence the integrated intensity can be approximated by

$$I_{CO} \sim T \, \Delta v \tag{3.42}$$

[†] ^{12}C^{16}O is the common isotopic variant of CO. Additional information can sometimes be obtained by also observing the ^{13}C^{16}O, ^{12}C^{17}O and ^{12}C^{18}O variants, which are less abundant, and therefore tend to be less optically thick.

where Δv is the width of the line. Δv is usually taken to be the width of the line at an intensity level equal to half of the line's maximum intensity. This 'full width at half maximum' (FWHM) is a commonly encountered observable parameter.

If we assume that the emitting cloud is such that its internal velocity dispersion is sufficient to support it against self-gravity, then this requires that $\Delta v \simeq (GM/R)^{1/2}$ (see Chapter 4, Section 4.4), and so

$$I_{CO} \sim T \left(\frac{GM}{R} \right)^{1/2}, \tag{3.43}$$

where we have assumed that the cloud is approximately spherical with mass M and radius R.

3.3.3 The CO to H$_2$ conversion factor

What we really want to know is the total amount of matter in the cloud producing the observed CO emission, and most of the matter is in the form of molecular hydrogen. Therefore we would like to be able to relate the column density of molecular hydrogen (N_{H_2}) to the integrated intensity of the CO line emission (I_{CO}). We call this ratio $\mathcal{X}$. If molecular hydrogen is the dominant form of hydrogen in the cloud (and we can usually safely assume this, if the cloud is detected in CO), then the mean column density of molecular hydrogen through the cloud is given by

$$N_{H_2} = \frac{(0.7 M / 2 m_p)}{\pi R^2}, \tag{3.44}$$

where m_p is the mass of a proton. Here the quantity in brackets is the total number of hydrogen molecules in the cloud, i.e. 70% of the cloud's total mass divided by the mass of a hydrogen molecule.

Combining this with equation 3.43, we obtain

$$\mathcal{X} \equiv \frac{N_{H_2}}{I_{CO}} \sim \frac{0.7}{2\pi m_p T} \left(\frac{M}{GR^3} \right)^{1/2}$$

$$\sim \frac{4 \times 10^{24} \text{m}^{-2}}{\text{K(km s}^{-1})} \left(\frac{T}{10 \text{ K}} \right)^{-1} \left(\frac{n_{H_2}}{10^9 \text{ m}^{-3}} \right)^{1/2}. \tag{3.45}$$

Since the temperatures in molecular clouds observed in the CO line are typically $T \sim 10$ K, and the mean densities are typically $n \sim 10^9$ m^{-3}, the ratio of N_{H_2} to I_{CO} should not vary much from one cloud to another. Thus I_{CO} is a good tracer of the total amount of molecular gas, in the sense that it is approximately proportional to the total column density of gas, provided that the CO line remains optically thin.

We have presented here a theoretical treatment. In reality, the approximate correlation between I_{CO} and N_{H_2} was established observationally, and the theory sketched above was then offered as an explanation for the observed correlation.

We note that in order to obtain an estimate for the mass of a cloud, we also need to know its distance D. Then the angular size of the cloud, Ω_c, gives its cross-sectional area, $A = \Omega_c D^2$, and the mass of the cloud M_{cl} is given by

$$N_{H_2} = \frac{(0.7 M_{cl}/2m_p)}{A}. \tag{3.46}$$

Rearranging gives

$$M_{cl} = A N_{H_2}(2m_p/0.7) = \Omega_c D^2 I_{CO}(2m_p/0.7)\chi. \tag{3.47}$$

In fact, any molecular species X can be used to infer the mass of a cloud from the measured intensity I_X of that species, provided the mass ratio is known between species X and molecular hydrogen H_2.

3.4 Line shapes and the motion of the gas

If species X has two energy levels i and j, with

$$E_j - E_i = h\nu_0, \tag{3.48}$$

an absorption line is produced by the reaction

$$X_i + \gamma_\nu \longrightarrow X_j. \tag{3.49}$$

The cross-section $\sigma(\nu)$ presented by a particle of species X_i to a photon γ_ν of frequency ν is

$$\sigma(\nu) = \sigma_0 \, \phi(\nu - \nu_0), \tag{3.50}$$

where σ_0 is the integrated absorption cross-section, given by

$$\sigma_0 = \frac{h\nu_0 B_{ij}}{c} = \frac{g_j c^2 A_{ji}}{8\pi g_i \nu_0^2}, \tag{3.51}$$

and $\phi(\nu - \nu_0)$ is again the profile function (see Section 2.9.4). In other words, a particle of species X in level i has a fixed amount of absorbing power measured by the integrated absorption cross-section σ_0. The profile function $\phi(\nu - \nu_0)$ simply determines whether this absorbing power is spread thinly over a wide range of frequencies or concentrated intensively in a narrow range. Hence it determines the shape of the spectral line.

3.4.1 Line broadening

There are two main factors determining the form of the profile function $\phi(\nu - \nu_0)$, known as natural broadening and Doppler broadening.

Natural broadening arises because the emitting particles have a finite life-time in the upper level of the transition, and so the energy of the upper level is uncertain, due to the uncertainty principle. At low densities, the mean life-time in the upper level is given by

$$\Delta t \sim A_{ji}^{-1}, \tag{3.52}$$

where A_{ji} is the Einstein A-coefficient for spontaneous radiative de-excitation. The corresponding uncertainty in the energy of the upper level is given by the uncertainty principle

$$\Delta E \sim \frac{h A_{ji}}{4\pi}, \tag{3.53}$$

and this translates into an uncertainty in the frequency of the transition

$$\Delta \nu_{\text{N}} \sim \frac{A_{ji}}{4\pi}, \tag{3.54}$$

where $\Delta \nu_{\text{N}}$ is the linewidth due to natural broadening.

When the process is analysed properly, the natural line profile is found to adopt a Lorentzian form

$$\phi_{\text{N}}(\nu - \nu_0) = \frac{\Delta \nu_{\text{N}}}{\pi \left[\Delta \nu_{\text{N}}^2 + (\nu - \nu_0)^2 \right]}. \tag{3.55}$$

A large A_{ji} makes for a large $\Delta \nu_{\text{N}}$, and hence for a broad, flat profile function $\phi_{\text{N}}(\nu - \nu_0)$.

If the density is sufficiently high that population is removed from the upper level by collisional de-excitation more rapidly than by spontaneous radiative de-excitation, the life-time in the upper level is significantly reduced and the line is broadened further. This is called pressure broadening. Pressure broadening is seldom important in interpreting observations of the interstellar medium, but can be very important in analysing stellar spectra, because of the higher densities that occur in stellar atmospheres.

3.4.2 The Doppler effect

An absorption line can be both shifted and broadened by the Doppler effect. To separate the two effects, we divide the radial velocity u_{rad} of an individual absorbing particle into a systematic part u_0, which is due to the bulk motion of the cloud in which the particle resides, and a random part u_1, which is due to the random motion of the individual particle relative to the centre of mass of the cloud

$$u_{\text{rad}} = u_0 + u_1. \tag{3.56}$$

The bulk radial velocity of the cloud (u_0) shifts the line centre to

$$\nu_0 = \nu_{\text{rest}} \left[1 - \frac{u_0}{c} \right], \tag{3.57}$$

where ν_{rest} is the rest frequency of the line. This is why discrete clouds having different bulk radial velocities should produce groups of lines with different central frequencies.

The random motion of the particles relative to the centre of mass of the cloud is characterised by a velocity dispersion

$$\Delta u_D = \left[\frac{kT}{m_X} + \Delta u_T^2 \right]^{1/2}. \tag{3.58}$$

Δu_D is compounded by thermal motions at a level of $\sim [kT/m_X]^{1/2}$, where m_X is the mass of a particle of species X, and turbulent motions at a level of $\sim \Delta u_T$. By turbulent motions we mean random bulk motions involving fluid elements which are microscopic from the perspective of the whole cloud, but macroscopic from the perspective of the individual gas particles. For simplicity, we shall not discuss turbulent motions further here and we shall assume

$$\Delta u_D \longrightarrow \left[\frac{kT}{m_X} \right]^{1/2}, \tag{3.59}$$

although in many clouds turbulent motions are the dominant source of velocity dispersion, and we shall return to a discussion of turbulence in the next chapter.

Since the thermal motions have a Maxwellian distribution, the distribution of radial velocities is Gaussian

$$n_{X,u_1} \, du_1 = \frac{1}{(2\pi)^{1/2} \, \Delta u_D} \exp \left[-\frac{(u_1 - u_0)^2}{2 \, \Delta u_D^2} \right] \, du_1. \tag{3.60}$$

Hence the Doppler line profile function, $\phi_D(\nu - \nu_0)$, is also Gaussian

$$\phi_D(\nu - \nu_0) = \frac{1}{(2\pi)^{1/2} \, \Delta \nu_D} \exp \left[-\frac{(\nu_1 - \nu_0)^2}{2 \, \Delta \nu_D^2} \right], \tag{3.61}$$

where $\Delta \nu_D$ is the frequency broadening due to the Doppler effect, given by

$$\Delta \nu_D = \frac{\nu_0 \Delta u_D}{c}. \tag{3.62}$$

From equations 3.59 and 3.62, we see that a high temperature and/or a low particle-mass of the relevant species (m_X) produce a large Δu_D and $\Delta \nu_D$, and hence a broad, flat profile function $\phi_D(\nu - \nu_0)$.

3.4.3 Convolving line profiles

To obtain the overall combined profile function, $\phi_C(\nu - \nu_0)$, we must convolve the natural and Doppler profile functions, to obtain

$$\phi_C(\nu - \nu_0) \, d\nu = \int_{\nu'=0}^{\nu'=\infty} \phi_N(\nu - \nu') \, \phi_D(\nu' - \nu_0) \, d\nu' \, d\nu. \tag{3.63}$$

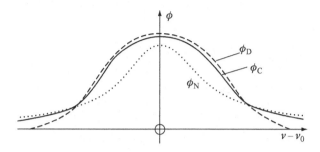

Fig. 3.7. Convolution of the natural profile, ϕ_N (dotted curve), with the Doppler profile, ϕ_D (dashed curve), to produce the overall profile ϕ_C (full curve).

This is a good example of the use of convolution, so we should try to understand the underlying principles.

The combined profile function $\phi_C(v - v_0)\,dv$ measures the fraction of the (fixed) total absorbing power of a particle of species X in level i which falls in the frequency range $(v, v + dv)$. Absorbing power arrives at this frequency range in two steps. The first step, with probability $\phi_D(v' - v_0)\,dv'$, Doppler-shifts the centre of the natural line into the intermediate frequency range $(v', v' + dv')$. The second step, with probability $\phi_N(v - v')\,dv$, delivers absorbing power into the frequency range of interest $(v, v + dv)$ by natural broadening. The location of the intermediate frequency range $(v', v' + dv')$ does not concern us, and so we add up all the different possibilities by integrating the product of the probabilities

$$\phi_D(v' - v_0)\,dv' \times \phi_N(v - v')\,dv$$

over v', which gives equation 3.63.

Usually, the natural width is significantly less than the Doppler width

$$\Delta v_N \ll \Delta v_D, \tag{3.64}$$

in which case the core (central part) of the combined profile is very close to the Doppler profile, and the wings (outer part) of the combined profile are very close to the natural profile, as illustrated in Figure 3.7. There is an intermediate part around

$$|v - v_0| \sim \Delta v_C > \Delta v_D, \tag{3.65}$$

where the switch-over is located.

This occurs because for small arguments $|v - v'|$, i.e. near the centre, the natural profile is much more sharply peaked than the Doppler profile, and so the convolution integral approximates to convolving the Doppler profile with a delta function, which would leave the Doppler profile unchanged. Conversely, for large arguments $|v' - v_0|$, i.e. far from the centre, the Doppler profile falls off more steeply than the natural profile, and so here the convolution integral approximates to convolving the natural profile with a delta function. Careful study of the

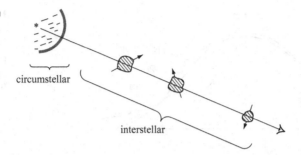

shapes of spectral lines can thus tell us about the kinematics of the gas in the ISM.

3.5 Absorption lines – searchlights through the ISM

Yet another powerful technique for studying the ISM is to use absorption lines. In this method we use distant luminous stars or other bright sources (e.g. active galactic nuclei) to act as beacons shining through the ISM between themselves and us – see Figure 3.8. The matter in the ISM between us and the distant star then absorbs light from the star at specific frequencies, allowing us to obtain information on the constitution of the ISM. However, we must be careful to differentiate between matter close to a star, which we call 'circumstellar', and matter spread between the stars, which we call 'interstellar'.

3.5.1 Selection effects

In order for there to be sufficient interstellar matter in front of a star to produce a measurable absorption line in its spectrum, the star must be quite distant, and therefore it must be quite luminous. This introduces a selection effect. Luminous stars, by their nature, tend to have a strong influence on their immediate surroundings, ionising the surounding gas and blowing off powerful stellar winds, which produce stellar-wind bubbles by sweeping up the surrounding gas into a dense shell.

So some of the absorption lines we see in the stellar spectrum will be due to ionised gas, stellar-wind bubbles, and the dense shells of gas which they sweep up. This gas is not representative of the interstellar medium at large, but of the immediate surroundings of a luminous star. In other words, it is circumstellar rather than interstellar. It is not always straightforward to distinguish absorption lines due to circumstellar gas from those due to interstellar gas.

Of course, this problem only arises because we do not normally have any direct way of ascertaining how the material producing an absorption line is distributed along the line of sight to the star. In principle, it could be concentrated in a thin layer anywhere along the line of sight, or it could be spread out uniformly between the observer and the star.

An additional selection effect arises because luminous stars are relatively rare. Therefore we can only measure absorption lines along a few particular lines of sight. Moreover, if we are interested in the cool components of the interstellar medium, which are concentrated in a thin layer near the Galactic midplane, we must observe stars at low Galactic latitude, otherwise only a small part of the line of sight to the star will intercept this layer.

There are other effects which can alter our column density estimates of the interstellar medium, and which can be difficult to account for. One such effect is that we do not know what fraction of a given element is in the gaseous phase and what fraction is in the solid phase, i.e. dust grains. This is especially true for elements such as Si and Ca. This effect could be dependent upon environment in ways we don't understand, and hence could introduce unknown selection effects.

The overall result is that deducing reliable information about the general interstellar medium from such observations is a painstaking business, and in the end rather model-dependent.

3.5.2 Circumstellar and interstellar lines

We now list the factors which point to some of the absorption lines seen in stellar spectra being produced in the interstellar medium at large, rather than in the stellar atmosphere or in a strictly circumstellar region.

Certain absorption lines are seen in the spectra of distant stars, but are not seen in the spectra of otherwise very similar nearby stars. Certain absorption lines seen in stellar spectra are due to species which are not stable under the conditions occurring in a stellar atmosphere. Usually they are species which would be ionised or dissociated at the temperatures and densities of a typical stellar atmosphere.

Certain absorption lines seen in stellar spectra are too narrow to have been produced in a stellar atmosphere. If T_* is the surface temperature of the star, m_X is the mass of a single particle of the absorbing species, and ν_0 is the frequency at the centre of the line, then the width of an absorption line produced in a stellar atmosphere should be

$$\Delta \nu \gtrsim (8 \ln[2])^{1/2} \frac{\nu_0}{c} \left(\frac{kT_*}{m_X} \right)^{1/2}, \qquad (3.66)$$

where the inequality arises if there is additional broadening of the line over and above thermal broadening (i.e. turbulent or pressure broadening).

The central frequency ν_0 of an absorption line may indicate that the absorbing gas has a different radial velocity from the background star. The existence of interstellar gas was first unambiguously confirmed from observations of a spectroscopic binary. The central frequencies of the absorption lines produced in the stellar atmosphere shifted with a regular period, as the star orbited its companion, sometimes moving away from the observer and sometimes towards. The central frequencies of the absorption lines produced in the intervening interstellar medium were constant.

Often close groups of absorption lines are observed, with all the lines in a group being attributable to the same transition in the same species. The inference is that the absorbing particles are not distributed uniformly along the line of sight to the background star, but are concentrated in discrete clouds having different bulk radial velocities.

The number of lines in a group tends to be larger for more distant stars. The inference is that, on average, the number of intervening clouds increases with the distance to the background star. This fact is particularly important in distinguishing lines produced in the general interstellar medium from lines produced in circumstellar material. All of these factors help to identify which lines are due to absorption in the ISM.

3.5.3 Equivalent width of a line

Observed absorption lines are often noisy and confused, so that it is not possible to fit the shape of the line in detail. The equivalent width of an absorption line is a measure of its total strength that is independent of the detailed shape, and therefore can be determined accurately, even for lines which are quite noisy and confused.

Figure 3.9(a) illustrates a fictitious group of five absorption lines, all due to the same transition, and thereby presumably attributable to five different clouds along the line of sight. The plotted quantities are the observed (measured) monochromatic flux, F_λ^{obs}, where

$$F_\lambda^{obs} \sim \frac{L_\lambda^*}{4\pi D^2} \exp\left[-\tau_\lambda\right], \qquad (3.67)$$

against the wavelength λ. The lines are noisy, in the sense that the variation of F_λ^{obs} with λ is not smooth. The lines are also confused, in the sense that they overlap. The first thing to do is to smooth the observed flux to produce a smoothed flux F_λ^{sm}

$$F_\lambda^{obs} \longrightarrow F_\lambda^{sm} = \frac{L_\lambda^*}{4\pi D^2} \exp\left[-\tau_\lambda\right]. \qquad (3.68)$$

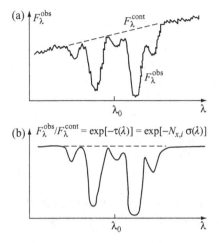

Fig. 3.9. (a) A group of absorption lines, all due to the same transition, but in different clouds along the line of sight to the background star. (b) The same set of lines after smoothing and normalising to the background continuum.

This is normally done numerically, taking care to choose a smoothing algorithm which effectively removes the noise without destroying the signal.

Next one estimates the background continuum flux, $F_\lambda^{\mathrm{cont}}$, which would have been measured if there were no interstellar absorption

$$F_\lambda^{\mathrm{cont}} = \frac{L_\lambda^*}{4\pi D^2}. \tag{3.69}$$

This is done by choosing points on either side of the absorption lines where there appears to be no absorption, and interpolating across the absorption lines by fitting them with a low-order polynomial. Again this is normally done numerically.

Finally one divides the smoothed flux by the continuum flux, to obtain the normalised flux f_λ

$$f_\lambda \equiv \frac{F_\lambda^{\mathrm{obs}}}{F_\lambda^{\mathrm{cont}}} = \exp\left[-\tau_\lambda\right] = \exp\left[-N_{X,i}\,\sigma(\lambda)\right], \tag{3.70}$$

where $N_{X,i}$ is the column density of absorbing particles (i.e. particles of species X in the lower level i), and $\sigma(\lambda)$ is the cross-section presented by a single particle of species X in level i to photons of wavelength λ. After smoothing and normalisation, the plot looks like that on Figure 3.9(b).

Figure 3.10 illustrates three types of absorption line on a plot of normalised flux f_λ against wavelength λ. The first line is unsaturated. In other words, even at the centre of the line there is still measurable flux. The second line is beginning to saturate – right at the centre the flux is too weak to measure accurately. The third line is strongly saturated – there is a wide band of wavelengths over which the flux is too weak to measure.

The equivalent width W_λ of an absorption line is the width of a rectangle which has the same area as that enclosed by the line on a

Fig. 3.10. Three different types of absorption line. From left to right, unsaturated, beginning to saturate, strongly saturated.

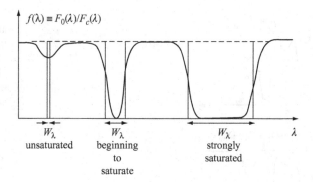

plot of normalised flux against wavelength. This is illustrated by the rectangular areas on Figure 3.10. Mathematically this is given by

$$W_\lambda = \int_{\text{line}} \{1 - f_\lambda\}\, d\lambda$$

$$= \int_{\text{line}} \left\{1 - \exp\left[-N_{X,i}\sigma(\lambda)\right]\right\} d\lambda$$

$$= \int_{\text{line}} \left\{1 - \exp\left[-N_{X,i}\sigma_0\phi(\lambda)\right]\right\} d\lambda. \tag{3.71}$$

Since the equivalent width W_λ depends on the column density of absorbing particles $N_{X,i}$, and on the radial velocity dispersion Δu_D (via the profile function $\phi(\lambda)$), we have

$$W_\lambda \longrightarrow W_\lambda(\Delta u_D, N_{X,i}).$$

Since the normalized flux is dimensionless, the units of the equivalent width – as defined here – are the units of the *abscissa*, i.e. the units of wavelength. The convention is to measure equivalent widths in wavelength units, because the equivalent widths of optical lines are often of order 0.1 nm. Unfortunately, most other aspects of radiation and radiative transfer are normally calculated and described in terms of frequency.

The equivalent width of a spectral line should not be confused with other measures of linewidth that we shall use. For example, the full width at half maximum (FWHM) is often used by astronomers to describe the width of a spectral line. The half-maximum points of a line are the positions at which the line intensity has fallen to half of its peak intensity. The FWHM is simply the width of the line between the two half-maximum points, expressed in some appropriate units, such as a difference in frequency $\Delta \nu$, or wavelength $\Delta\lambda$ ($= c\Delta\nu/\nu^2$). Alternatively we can use the Doppler formula to convert frequency width to velocity width Δv

$$\Delta v = \frac{c\Delta\nu}{\nu}. \tag{3.72}$$

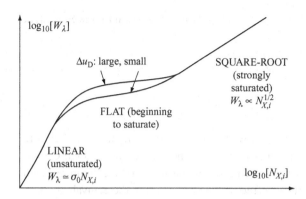

Fig. 3.11. A curve of growth.

In this way we can convert Δv into velocity units, and hence express the FWHM in terms of velocity Δv. For a Gaussian line shape the FWHM can be related to the standard deviation σ of the Gaussian by the standard formula

$$\Delta v = (8 \ln[2])^{1/2}\sigma. \tag{3.73}$$

In this case the standard deviation σ of the line is more commonly referred to as the velocity dispersion. We will return to the use of the velocity dispersion in Chapter 4.

3.6 The curve of growth

A plot of equivalent width W_λ against the column density of absorbing particles $N_{X,i}$ is called a 'curve of growth'. It is normally plotted on a log-log graph. A curve of growth divides into three sections (see Figure 3.11).

For small column densities, the absorption line is unsaturated and the curve of growth is linear

$$W_\lambda \propto N_{X,i}. \tag{3.74}$$

For intermediate column densities, the absorption line is beginning to saturate and the curve of growth is flat – i.e. W_λ is only weakly dependent on $N_{X,i}$. In addition, this is where the dependence on velocity dispersion, Δu_D, is strongest.

For large column densities, the absorption line is strongly saturated and the curve of growth has a square-root form

$$W_\lambda \propto N_{X,i}^{1/2}. \tag{3.75}$$

3.6.1 Low optical depth

When the column density of absorbing particles is sufficiently small, the optical depth is low even at the line centre, and the argument of the exponential in equation (3.71) is much less than unity. So we can use the approximation $1 - e^{-\epsilon} \simeq \epsilon$ to simplify the expression for the equivalent width

$$W_\lambda = \int_{\text{line}} \left\{ 1 - \exp\left[-N_{X,i}\sigma_0\phi(\lambda) \right] \right\} d\lambda$$

$$\simeq \int_{\text{line}} N_{X,i}\sigma_0\phi(\lambda)\, d\lambda = N_{X,i}\sigma_0. \tag{3.76}$$

What this means is that when the line is optically thin and therefore unsaturated, the absorbing particles do not shield one another. Consequently, if we add one more absorbing particle to the column between the observer and the star, the resulting increase in the equivalent width reflects the full absorbing power of the particle, as measured by the integrated absorption cross-section σ_0, hence the curve is linear

$$\frac{dW_\lambda}{dN_{X,i}} = \sigma_0. \tag{3.77}$$

Note that if we work in wavelength units, the integrated absorption cross-section is different from the one defined in equation 3.51. In frequency units, the cross-section presented to a photon of frequency ν is given by $\sigma(\nu) = \sigma_0\phi(\nu - \nu_0)$. Remember that $\sigma(\nu)$ has the dimensions of area, and ϕ has the dimensions of one over frequency (in order to ensure normalization – see equation 2.69), so σ_0 has the dimensions of area $\times$ frequency – e.g. m^2 Hz.

In wavelength units, the cross-section presented to a photon of wavelength λ is given by $\sigma(\lambda) = \sigma_0\phi(\lambda - \lambda_0)$. Here $\sigma(\lambda)$ again has the dimensions of area, but now ϕ has the dimensions of one over wavelength (in order to ensure normalisation), so σ_0 now has the dimensions of area $\times$ wavelength – e.g. m^2 nm – and equation 3.51 must be replaced by

$$\sigma_0 = \frac{h\lambda_0 B_{ij}}{c} = \frac{g_j \lambda_0^4 A_{ji}}{8\pi g_i c}. \tag{3.78}$$

3.6.2 Intermediate optical depth

When the column density of absorbing particles is intermediate, the optical depth is only significant near the line centre. In the wings of the line, the optical depth is low. If we add a new particle to the column between the observer and the star, the particle is shielded from the radiation near the line centre and presents a very small cross-section to the radiation in the line wings.

Hence the effect of the additional particle on the equivalent width, $dW_\lambda/dN_{X,i}$, is very small and the curve of growth is flat (see Figure 3.11). This means that estimating $N_{X,i}$ from W_λ is difficult, because a small uncertainty in W_λ converts into a large uncertainty in $N_{X,i}$.

Furthermore, the flat portion of the curve depends strongly on the radial velocity dispersion Δu_D, and this increases the uncertainty further. In order to overcome this problem, it is necessary to measure at least two absorption lines, preferably from the same level of the same species. We can then estimate the velocity dispersion, Δu_D, as well as the column density.

3.6.3 High optical depth

When the column density of absorbing particles is large, the optical depth is large over a wide range of frequencies and the line is strongly saturated. If we add a new particle to the column between the observer and the star, the only radiation which is left for the particle to absorb is so far out in the wings of the line that the profile here approximates to the natural form

$$\phi_N(\lambda - \lambda_0) = \frac{\Delta\lambda_N}{\pi\left[\Delta\lambda_N^2 + (\lambda - \lambda_0)^2\right]} \simeq \frac{\Delta\lambda_N}{\pi(\lambda - \lambda_0)^2}. \tag{3.79}$$

It is straightforward to show that in this limit

$$\frac{dW_\lambda}{dN_{X,i}} \propto N_{X,i}^{-1/2}, \tag{3.80}$$

or in other words, the more absorbing particles there are in the column, the smaller the effect of adding one more absorbing particle. Therefore

$$W_\lambda \propto N_{X,i}^{1/2}. \tag{3.81}$$

The foregoing shows us that an understanding of the growth of spectral linewidths with increasing column density is essential to interpreting measurements of spectral lines.

3.7 The use of absorption lines

As an example of the use of absorption lines, suppose that we have measured W_λ for the H and K lines[†] of Ca^+ in the spectrum of a distant star. What we really need to estimate is the column density N of hydrogen in all forms on this line of sight, and hence the mean volume density n of hydrogen in all forms. To do this there are several steps we must carry out.

- Perform a curve of growth analysis to determine the column density of absorbing particles $N_{Ca^+,i}$.

[†] These lines have $\lambda_H = 396.85$ and $\lambda_K = 393.37$ nm.

- Consider the excitation balance to estimate what fraction of all the Ca^+ is in level i and correct for the rest to obtain the total column density of Ca^+, N_{Ca^+}. Since level i is the ground state of Ca^+, most of the Ca^+ is probably in this level, so this is not usually a significant correction.
- Consider the ionisation balance to estimate what fraction of the gas-phase calcium is in the singly ionised state and correct for the rest to obtain the total column density of gas-phase calcium, N_{Ca}^{gas}. This is a much more difficult and model-dependent correction.
- Consider depletion to estimate what fraction of the calcium is locked up in dust grains (and therefore absent from the gas phase), and correct for it to obtain the total column density of calcium N_{Ca}^{tot}. Again this is difficult to do. We don't know what fraction of the Ca is locked away in dust grains.
- Adopt the cosmic abundance of calcium relative to hydrogen and hence convert N_{Ca}^{tot} into N (the column density of hydrogen in all forms).
- Consider how uncertain the final result is, in particular how critically it depends on the assumption (implicit in the above analysis) that conditions in the interstellar medium are completely uniform along the line of sight.

This shows how the study of absorption lines can be used to estimate the mass of gas in the ISM – and how uncertain such estimates can be.

Recommended further reading

We recommend the following texts to the student for further reading on the topics presented in this chapter.

Banwell, C. N. and McCash, E. M. (1994). *Fundamentals of Molecular Spectroscopy*, 4th edn. New York: McGraw-Hill.

Dyson, J. E. and Williams, D. A. (1997). *The Physics of the Interstellar Medium*, 2nd edn. Bristol: Institute of Physics Press.

Spitzer, L., Jr (1978). *Physical Processes in the Interstellar Medium*. New York: Wiley.

Chapter 4
Molecular clouds – the sites of star formation

4.1 The equation of state

Star formation takes place predominantly in large clouds of gas and dust which inhabit the interstellar medium of our Galaxy and of other galaxies. Therefore, to understand the process of star formation, we must first understand the physical processes which take place in these clouds. In this chapter we will look at the stability of molecular clouds and begin to assess what might be needed to make such clouds collapse under gravity and form stars.

4.1.1 The ideal gas approximation

The behaviour of a gas under different physical conditions is described by the equation of state of the gas. The simplest assumption we can adopt is that a molecular cloud behaves as an ideal gas at constant temperature, obeying the equation

$$PV = \mathcal{N}kT, \tag{4.1}$$

where P is the pressure, V is the volume, $\mathcal{N}$ is the total number of molecules, k is the Boltzmann constant, and T is the temperature. If we eliminate $\mathcal{N}$ and V using the equation for the density

$$\rho = \frac{\mathcal{N}\mu m_{\mathrm{H}}}{V}, \tag{4.2}$$

where m_{H} is the mass of a hydrogen atom and μ is the mean molecular weight of the gas (which would be two if the cloud were entirely composed of molecular hydrogen), then we obtain

$$P = \frac{\rho k T}{\mu m_{\mathrm{H}}},$$ (4.3)

which relates the pressure to the temperature and density of the gas. It is often convenient when treating astrophysical problems relating to molecular clouds to assume that the gas temperature and chemical composition remain constant. This is known as the isothermal assumption, under which equation 4.3 becomes

$$\frac{P}{\rho} = \frac{kT}{\mu m_{\mathrm{H}}} = a_0^2 = \text{constant},$$ (4.4)

where a_0 is the isothermal sound speed in the gas at temperature T.

4.1.2 Adiabatic equation of state

The next simplest assumption we can adopt is the adiabatic equation of state

$$p\rho^{-\gamma} = K,$$ (4.5)

where K is constant and γ is the ratio of specific heats at constant pressure, c_p, and volume, c_v, respectively: $\gamma = c_p/c_v$. We will derive more general forms of the equation of state later in this chapter.

4.2 Fluid mechanics of molecular clouds

The material in molecular clouds is rarely stationary, and we must take this into account. To achieve this, we use the basic equations of fluid mechanics, which we first derive. In general, there are two methods commonly used to model gas flow. One method is to use a fixed set of coordinates in space and calculate the parameters of the gas as it flows through the coordinate frame. This is known as the Eulerian method. An alternative is to choose a set of coordinates fixed to a particle of the gas, moving with that particle, and to calculate the varying parameters in that coordinate frame (referred to as comoving coordinates). This technique is known as the Lagrangian method.

4.2.1 The continuity equation

Consider a non-viscous fluid in which the density and velocity are functions of position and time, i.e. $\rho = \rho(\mathbf{r}, t)$ and $v = v(\mathbf{r}, t)$. Now focus on an arbitrary volume V contained by a closed surface S, as shown in Figure 4.1. V and S are fixed in space, so we are here adopting the

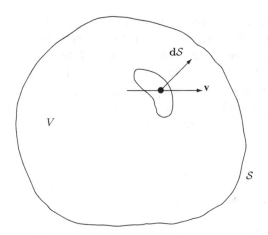

Fig. 4.1. An arbitrary volume V, contained by a closed surface S. $\mathbf{d}S$ is an infinitesimal vector area element, whose direction by convention is normal to the surface S and out of V. At the position of $\mathbf{d}S$ the density of the fluid is ρ and its velocity is $\mathbf{v}$, so matter flows out of V across $\mathbf{d}S$ at a rate $\rho\mathbf{v}.\mathbf{d}S$.

Eulerian viewpoint. The mass flowing out of V through the element of area $\mathbf{d}S$ is given by

$$\rho\mathbf{v}.\mathbf{d}S,$$

and so the net rate at which mass flows out of V through S is given by

$$\oint_S \rho\mathbf{v}.\mathbf{d}S = \int_V \nabla.(\rho\mathbf{v})dV,$$

where we have obtained the right-hand side by invoking Gauss's divergence theorem.

The rate at which the mass in V decreases is given by

$$-\frac{\partial}{\partial t}\left(\int_V \rho dV\right) = \int_V \left(-\frac{\partial \rho}{\partial t}\right)dV, \tag{4.6}$$

where we can take $\partial/\partial t$ inside the integral because V is fixed in space. Obviously, the rate at which the mass in V decreases (equation 4.6) must equal the rate at which mass flows out of V across S, so

$$\int_V \left[\frac{\partial \rho}{\partial t} + \nabla.(\rho\mathbf{v})\right]dV = 0, \tag{4.7}$$

and since the volume V is arbitrary, it follows that

$$\frac{\partial \rho}{\partial t} + \nabla.(\rho\mathbf{v}) = 0 \tag{4.8}$$

everywhere. This is known as the continuity equation, and is one of the fundamental equations of fluid mechanics. We can expand this equation and rewrite it as

$$\frac{\partial \rho}{\partial t} + \rho\nabla.\mathbf{v} + \mathbf{v}.\nabla\rho = 0. \tag{4.9}$$

The significance of this equation is that it represents the principle of conservation of mass for a fluid flow. We shall return to this equation shortly.

We stress that $\partial\rho/\partial t$ is the Eulerian time derivative of the density (i.e. the rate of change of density at a fixed point in space). If we want the Lagrangian time derivative of the density (i.e. the rate of change of density moving with the fluid) we must include the contribution due to the displacement, $\mathbf{dr} = \mathbf{v}dt$, which occurs during the time interval dt. The net density change is

$$d\rho = \frac{\partial\rho}{\partial t}dt + \mathbf{dr}.\nabla\rho, \tag{4.10}$$

and hence the Lagrangian time derivative of the density is

$$\frac{d\rho}{dt} = \frac{\partial\rho}{\partial t}dt + \mathbf{v}.\nabla\rho = -\rho(\nabla.\mathbf{v}), \tag{4.11}$$

where the final expression is obtained by substituting from equation 4.9. $d\rho/dt$ is sometimes called the comoving time derivative of the density.

4.2.2 The equation of motion under pressure

Consider once again the volume of fluid V in Figure 4.1. If the non-viscous fluid in this volume has pressure $P(\mathbf{r}, t)$, then the total force acting on the volume is the sum of the external pressure on the surface. This is given by the surface integral

$$-\oint_S P d\mathcal{S}.$$

Transforming this into a volume integral, the net pressure force exerted on the arbitrary volume V is

$$-\oint_S P d\mathcal{S} = -\int_V \nabla P dV, \tag{4.12}$$

and hence the net pressure force per unit volume is simply $-\nabla P$.

The equation of motion of this volume can be derived by equating the force per unit volume with the mass per unit volume multiplied by its acceleration. This is simply Newton's third law. The mass per unit volume is defined as the density ρ, and the acceleration is the time derivative of the velocity $d\mathbf{v}/dt$. So we have

$$-\nabla P = \rho\frac{d\mathbf{v}}{dt}, \tag{4.13}$$

and hence

$$\frac{d\mathbf{v}}{dt} = -\frac{\nabla P}{\rho}. \tag{4.14}$$

Here $d\mathbf{v}/dt$ is the comoving acceleration of the fluid, so equation 4.14 is the Lagrangian formulation of the equation of motion.

The Eulerian formulation is obtained by substituting for $d\mathbf{v}/dt$ using

$$\frac{d\mathbf{v}}{dt} = \frac{\partial \mathbf{v}}{\partial t} + (\mathbf{v}.\nabla)\mathbf{v}. \tag{4.15}$$

Compare this with the Lagrangian time derivative of the density (equation 4.11). Substituting this into equation 4.14 gives

$$\frac{\partial \mathbf{v}}{\partial t} + (\mathbf{v}.\nabla)\mathbf{v} + \frac{\nabla P}{\rho} = 0. \tag{4.16}$$

This is sometimes referred to as Euler's equation. Its significance is that it is the equation of motion of an ideal fluid, acted on only by its own pressure.

If the fluid has significant viscosity, there are additional terms, and the Euler equation becomes the Navier–Stokes equation. However, in the interstellar medium the viscosity is very low, and therefore the extra terms can normally be omitted. The Navier–Stokes equation is the general equation of motion of a viscous incompressible fluid (an extra term must be included for compressible fluids).

4.2.3 Fluid motion under gravity

If the fluid is also in a gravitational field, then an extra term must be included in equation 4.14 to account for this. The force on unit volume due to a gravitational acceleration $\mathbf{g}$ is simply $\rho\mathbf{g}$, and so equation 4.14 becomes

$$\rho\frac{d\mathbf{v}}{dt} + \nabla P - \rho\mathbf{g} = 0, \tag{4.17}$$

and hence Euler's equation becomes

$$\frac{\partial \mathbf{v}}{\partial t} + (\mathbf{v}.\nabla)\mathbf{v} + \frac{\nabla P}{\rho} - \mathbf{g} = 0. \tag{4.18}$$

These are then the basic equations of fluid mechanics.

4.3 Gravitational instability

If a star is to form in a molecular cloud, then the cloud must become gravitationally unstable, and subsequently collapse. In this section we consider the question of a cloud's stability.

4.3.1 Uniform density medium

Consider an infinite static three-dimensional medium, with initial uniform density ρ_0 and uniform isothermal sound speed a_0. Now suppose that due to a random statistical fluctuation a portion of this medium becomes slightly more dense. For simplicity we assume that the portion is spherical, with radius r, and that it responds isothermally.

We wish to know whether the spherical portion continues to become denser and condenses out due to its self-gravity, or whether its internal pressure causes it to expand back to the same density as the surrounding medium. The answer depends upon the size of the spherical portion. For a large portion the self-gravity overcomes internal pressure, while for a small portion the internal pressure resists gravity.

To see this we write an equation for the radial excursions of the portion of gas in question. The outward acceleration due to its pressure is $\nabla P / \rho_0$. Since $\nabla P \sim P/r$ and $P = a_0^2 \rho_0$, we have

$$\frac{\nabla P}{\rho_0} \sim \frac{a_0^2}{r}. \tag{4.19}$$

The inward acceleration due to self-gravity is $-GM/r^2$, where M is the portion of gas under study. Since $M = r^3 \rho$ we have

$$\frac{-GM}{r^2} \sim -G\rho_0 r. \tag{4.20}$$

Thus the equation of motion controlling radial excursions is

$$\ddot{r} \sim \frac{a_0^2}{r} - G\rho_0 r. \tag{4.21}$$

For continued increasing density, and hence condensation, we require $\ddot{r} < 0$ and therefore

$$G\rho_0 \gtrsim \frac{a_0^2}{r^2}, \tag{4.22}$$

leading to the condition for condensation

$$r > r_J \sim \frac{a_0}{(G\rho_0)^{1/2}}. \tag{4.23}$$

r_J is called the Jeans length. This is the minimum initial radius for a spherical portion of a uniform medium (characterised by a_0 and ρ_0), which can condense out due to its own self-gravity.

4.3.2 The Jeans mass

There is an equivalent minimum initial mass, M, associated with the Jeans length, such that

$$M > M_J \sim \frac{4}{3}\pi\rho_0 r_J^3 \sim \frac{4\pi a_0^3}{3(G^3 \rho_0)^{1/2}}. \tag{4.24}$$

M_J is called the Jeans mass. It is the minimum mass of a spherical portion of a uniform medium which can condense out due to its own self-gravity. We note that M_J depends on a_0^3, and from equation 4.4 we see that a_0 depends on $T^{1/2}$. Hence the temperature dependence of the Jeans mass (assuming all other parameters are held constant) is given by

$$M_J \propto T^{3/2}. \tag{4.25}$$

To give the reader some feel for the orders of magnitude of the values of the Jeans length and mass encountered in studies of molecular clouds, we can insert some typical values into equation 4.24. The densest regions of molecular clouds in which no stars have yet formed are typically observed to have temperatures of around 20 K, densities of around 10^{11} hydrogen molecules per m^3, and a mean molecular weight of 2.3 (i.e. mainly molecular hydrogen).

For such a region we apply equation 4.23 and derive a Jeans length of about 0.05 parsec. Using equation 4.24 we find that the Jeans mass would be roughly 3 $M_\odot$, where $M_\odot$ represents the Solar mass ($\simeq 2 \times 10^{30}$ kg).

So if, for instance, we observed a uniform density region at 20 K which was 0.1 pc in diameter, and contained 5 $M_\odot$ of matter, we would note that this was greater than its Jeans mass. Therefore, we would say that it was Jeans-unstable, and we would predict that it was about to collapse to form a star.

4.3.3 Structure in molecular clouds

One problem with the Jeans picture is that no extended, uniform, static clouds in the earliest stage of fragmentation have ever been observed. All molecular clouds are inhomogeneous and clumped on all scales. So if the theoretical situation of a large, uniform density, static cloud actually occurs, it must be such a short-lived phase that we seldom have the chance to observe it. In addition, many clouds show supersonic motions, indicating that the sound crossing time is not the relevant time-scale for the passage of information across a cloud – one of the basic assumptions of the Jeans theory.

Figure 4.2 shows an image of the Ophiuchus molecular cloud region taken at a wavelength of 100 microns. Even a cursory glance at Figure 4.2 shows that this region is extremely inhomogeneous. Structure can be seen on all scales from the resolution of the telescope to the full size of the image. The picture clearly shows a situation which is far from the constant-density cloud envisaged by Jeans.

The approximately circular or elliptical features seen in the image are often referred to as dense molecular cloud cores, and it is in some of these regions that the highest column densities of gas are observed. The elongated features are usually referred to as filaments. The dense cloud

Fig. 4.2. Image of the molecular cloud in the constellation of Ophiuchus, taken at a wavelength of 100 microns. Structure in the form of cores and filaments can be seen on all scales within the image.

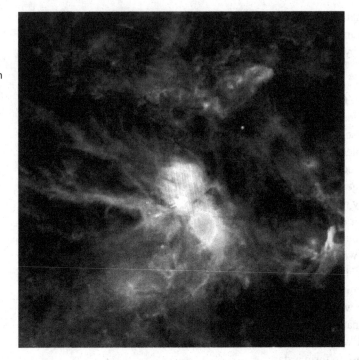

cores are the sites where star formation is believed to take place. Therefore, to deal with clouds of such complexity, we need a more detailed theoretical model than the Jeans mass of a uniform-density gas cloud.

We would also like to be able to answer questions such as whether the structures are symptomatic of cloud fragmentation; whether the observed gas motions are due to collapse, rotation, systematic or random motions; and what the dominant physical processes are which are causing the structure and motion which we observe.

4.4 The virial theorem

We now consider the energy balance in an isolated cloud, which is in equilibrium. We place no constraint on the density and velocity distributions. We look for a general theorem for the stability of such a cloud.

4.4.1 Cloud stability

We treat the cloud as an ensemble of particles, having mass m_i, position $\mathbf{r_i}$, and velocity $\mathbf{v_i}$, moving in their mutual gravitational field. The moment of inertia of the cloud $\mathcal{I}$ is given by

$$\mathcal{I} = \sum_i (m_i \mathbf{r_i} . \mathbf{r_i}).   \tag{4.26}$$

Then in the case of a molecular cloud in equilibrium, the time derivative of $\mathcal{I}$ must be zero. So we have

$$\dot{\mathcal{I}} = 2 \sum_i (m_i \mathbf{v_i}.\mathbf{r_i}) = 0, \tag{4.27}$$

since $\dot{m} = 0$, $\dot{\mathbf{r_i}} = \mathbf{v_i}$, and $\mathbf{v_i}.\mathbf{r_i} = \mathbf{r_i}.\mathbf{v_i}$. Furthermore, in equilibrium, the second time derivative must also be zero. So we have

$$\ddot{\mathcal{I}} = 2 \sum_i (m_i \dot{\mathbf{v_i}}.\mathbf{r_i}) + 2 \sum_i (m_i \mathbf{v_i}.\mathbf{v_i}) = 0. \tag{4.28}$$

Let us call the first term on the right-hand side of equation 4.28, $\ddot{\mathcal{I}}_1$. This can be rewritten, using Newton's second law, as

$$\ddot{\mathcal{I}}_1 = 2 \sum_i (m_i \dot{\mathbf{v_i}}.\mathbf{r_i}) = 2 \sum_i (\mathbf{F_i}.\mathbf{r_i}), \tag{4.29}$$

where $\mathbf{F_i} = m_i \dot{\mathbf{v_i}}$ is the force acting on the ith particle. For an isolated cloud, $\mathbf{F_i}$ represents the forces acting on i due to all the other particles j, i.e.

$$\mathbf{F_i} = \sum_{j \neq i} (\mathbf{F_{ij}}), \tag{4.30}$$

where $\mathbf{F_{ij}}$ is the force on particle i due to particle j. Thus

$$\ddot{\mathcal{I}}_1 = 2 \sum_i \sum_{j \neq i} (\mathbf{F_{ij}}.\mathbf{r_i}), \tag{4.31}$$

and, since the ordering of summation is immaterial,

$$\ddot{\mathcal{I}}_1 = \sum_i \sum_{j \neq i} (\mathbf{F_{ij}}.\mathbf{r_i} + \mathbf{F_{ji}}.\mathbf{r_j}). \tag{4.32}$$

Using Newton's third law, $\mathbf{F_{ij}} = -\mathbf{F_{ji}}$, this becomes

$$\ddot{\mathcal{I}}_1 = \sum_i \sum_{j \neq i} (\mathbf{F_{ij}}.[\mathbf{r_i} - \mathbf{r_j}]). \tag{4.33}$$

We can therefore neglect short-range forces with $\mathbf{r_i} - \mathbf{r_j} \ll 1$ and put

$$\mathbf{F_{ij}} = -\frac{G m_i m_j}{|\mathbf{r_i} - \mathbf{r_j}|^3} (\mathbf{r_i} - \mathbf{r_j}). \tag{4.34}$$

whence

$$\ddot{\mathcal{I}}_1 = \sum_i \sum_{j \neq i} \left(\frac{G m_i m_j}{|\mathbf{r_i} - \mathbf{r_j}|^3} [\mathbf{r_i} - \mathbf{r_j}] \right) = 2\Omega_G, \tag{4.35}$$

where Ω_G is the self-gravitational potential energy of the cloud. The factor of 2 arises because the summation in equation 4.35 counts the mutual potential energy of each pair of particles twice.

If we now call the second term on the right-hand side of equation 4.28, $\ddot{\mathcal{I}}_2$, this can be rewritten as

$$\ddot{\mathcal{I}}_2 = 4\mathcal{K} \tag{4.36}$$

where $\mathcal{K}$ is the net translational kinetic energy of the particles of the cloud, due both to random thermal motions, and to bulk motions (such as turbulence and rotation).

Combining these results we have

$$\frac{1}{2}\ddot{\mathcal{I}} = \Omega_G + 2\mathcal{K} = 0. \tag{4.37}$$

This relation is called the virial theorem. It is a necessary (but not sufficient) condition for equilibrium. It can be modified to include the effects of a magnetic field or external pressure by adding extra terms. For example, an external pressure P_{EXT} introduces a term $-3P_{\text{EXT}}V$, where V is the volume of the cloud, so that

$$\Omega_G + 2\mathcal{K} - 3P_{\text{EXT}}V = 0. \tag{4.38}$$

We now proceed to describe some of the consequences of the virial theorem.

4.4.2 The virial mass

For a spherical cloud of mass M, radius R, and velocity dispersion σ (see equation 3.73), we have

$$\mathcal{K} = \frac{1}{2}M\sigma^2, \tag{4.39}$$

and

$$\Omega_G \sim -\frac{GM^2}{R}. \tag{4.40}$$

Substituting these into the virial theorem, one obtains

$$\sigma^2 \sim \frac{GM}{R}. \tag{4.41}$$

This form of the virial theorem leads to the concept of a virial mass, M_{vir}, given by

$$M_{\text{vir}} \sim \frac{\sigma^2 R}{G}. \tag{4.42}$$

If the mass of a cloud is roughly equal to its virial mass, then it is close to virial equilibrium. If a cloud has a mass greater than its virial mass, then it will collapse unless supported by some other mechanism.

If a cloud has a mass less than its virial mass, then it is not gravitationally bound and will probably disperse under the action of its own internal motions, unless it is confined by an external pressure. Note that the virial mass is proportional to σ^2, so molecular clouds may be dispersed by stellar winds and other similar effects, once star formation has begun.

4.4.3 Theoretical core life-times

Consider a molecular cloud core whose mass significantly exceeds its virial mass. In the absence of any mechanism of support, the core will collapse under self-gravity. The time-scale which characterises this collapse is known as the free-fall time-scale, t_{ff}, and is given by

$$t_{ff} = \left(\frac{3\pi}{32G\rho} \right)^{1/2}. \tag{4.43}$$

Alternatively consider a molecular cloud core whose mass is much less than its virial mass. In the absence of external pressure the core will disperse. The dispersion time-scale is given by

$$t_{disp} \simeq \frac{R}{\sigma}, \tag{4.44}$$

where R is the core radius and σ is its internal velocity dispersion.

4.5 Observations of molecular clouds

We now introduce some observations and attempt to interpret these in terms of the foregoing theory.

4.5.1 Larson's scaling relations

Observations of molecular clouds have revealed approximate scaling relations between their masses, M, radii, R and internal velocity dispersions, σ.[†] We use radius here, and elsewhere, as a notional measure of the linear extent of a cloud.

These relations are strictly empirical, and appear to apply both to whole clouds, and to substructures within clouds, over a wide range of masses and environments. For example, Figure 4.3 shows that the statistical correlation between velocity dispersion and radius can be fit by

$$\sigma \propto R^{0.5}. \tag{4.45}$$

Another similar relation is seen for the cloud mass, such that

$$\sigma \propto M^{0.25}, \tag{4.46}$$

[†] Remember that σ is related to the FWHM velocity, Δv, of an observed spectral line, by $\sigma = \Delta v/(8 \ln 2)^{1/2}$.

Fig. 4.3. Graph of σ versus radius (here labelled S), on a log-log scale, for 273 different molecular clouds. The straight line illustrates one of the Larson relations.

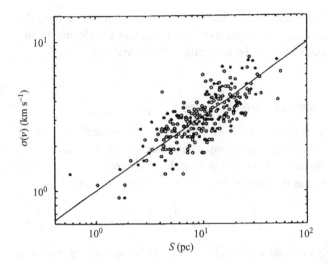

hence

$$M \propto R^2, \tag{4.47}$$

and thus we have a relation for density $\rho = M/R^3$, given by

$$\rho \propto R^{-1}. \tag{4.48}$$

All of the data appear to show that equations 4.45–4.48 provide a reasonably good fit to molecular clouds over a range of size-scales from roughly 0.05 pc to 100 pc.

One explanation for equation 4.46 could simply be that more massive clouds will tend to form more massive stars, and hence the massive stars will cause a greater disruption to the clouds in which they form, via their stellar winds. However, this explanation does not appear to be correct, since it holds equally true for clouds which have not yet begun to form stars, as for those in which the first star formation has begun.

Let us compare the relations derived from the data with the theoretical predictions of the virial theorem. From equations 4.45 and 4.46 we see that $R \propto \sigma^2$ and $M \propto \sigma^4$, hence

$$\frac{M}{R} \propto \sigma^2, \tag{4.49}$$

and therefore

$$\frac{M}{R\sigma^2} = \text{constant}. \tag{4.50}$$

From the virial theorem (equation 4.41), we have

$$\frac{M}{R\sigma^2} \sim \frac{1}{G}. \tag{4.51}$$

Thus the equivalence of equations 4.50 and 4.51 shows that the data for molecular clouds, such as those illustrated in Figure 4.3, appear to be at least dimensionally consistent with the virial theorem.

4.5.2 Cloud life-times from observations

We can estimate the life-times of the largest giant molecular clouds (GMCs) from the data illustrated in Figure 4.3, using equation 4.44. If we use $R = 100$ pc and its appropriate velocity dispersion, $\sigma \sim 10$ km s^{-1}, we derive a value for the dispersion time-scale t_{disp} of the order of 1.5×10^7 years. Thus, on this basis we would expect GMC life-times of typically fifteen million years. This is consistent with the crossing times of spiral arms within our Galaxy, which is the typical length of time for which we might expect a GMC to survive.

Alternatively, looking at the disruptive effects of star formation within molecular clouds, we find that the spread in age of the young stars observed in a newly formed cluster within a molecular cloud is typically around 10^7 years. Hence on the largest scales t_{disp} probably gives a reasonable estimate of molecular cloud life-times.

We can estimate the life-time of molecular cloud cores before they form stars using the relative statistics of cores with and without embedded young stars within them – although this method can only produce an order of magnitude estimate. The results of such surveys are that there are roughly equal numbers of cores both with and without embedded infrared sources. We can deduce therefore that the cores probably spend a roughly similar length of time before forming stars, as they subsequently spend with embedded stars in their centres.

The infrared sources can be mostly identified as T Tauri stars, which are pre-main-sequence stars whose evolution has been modelled (see Chapter 6) and consequently their life-times have been calculated. On the basis of this we can infer that the life-time of a core is of the order of a few million years before it forms a star in its centre.

4.5.3 Are cores in free-fall collapse?

We can compare the life-times of cores without stars (or starless cores) with their free-fall time-scales. The volume number density, $n(H_2)$, of a typical core is around 10^{10} H_2 molecules per m^3 ($\sim 4 \times 10^{-17}$ kg m^{-3}). Using equation 4.43 above yields a free-fall time-scale of $\sim 10^{13}$ s, or $\sim 3 \times 10^5$ years.

Hence the typical life-times of starless cores, inferred on the basis of the statistics of the numbers of cores observed, are roughly an order of magnitude longer than their free-fall life-times. Thus these cores cannot be undergoing free-fall collapse, and some physical process must be preventing them from collapsing. Processes which we have not yet considered include the effects of an interstellar magnetic field or of turbulence.

Measurements such as polarisation observations have shown that magnetic fields are to be found everywhere in the Galaxy, and clearly they will have an effect on the interstellar medium. This effect acts as an additional pressure resisting collapse, and the time taken for this pressure to dissipate is a few 10^6 years, consistent with starless core life-times. This is just one piece of evidence suggesting that magnetic fields play a role in regulating star formation in molecular cloud cores. We will return in the next chapter to the place occupied by starless cores in the evolutionary process of star formation.

4.6 Turbulence in molecular clouds

In this section we show evidence that turbulent velocity structure is probably present in molecular clouds. The stable flow of a fluid was discussed above. However, an unstable or turbulent flow is more complex. A fluid flow problem can typically be characterised by the Reynolds number, $\mathcal{R}$, of the flow. A critical Reynolds number exists for most flows, above which the flow becomes unstable to small perturbations, and turbulence develops.

When molecular clouds are observed in molecular line radiation, the line profiles cannot be explained by thermal broadening alone. There is additional broadening which must be attributable to bulk motions within the cloud, including turbulence.

4.6.1 Non-thermal linewidths

Spectral line profiles which are Gaussian in shape can often appear broader than they would be if the linewidth were due simply to thermal broadening. The thermal velocity dispersion of a spectral line emitted by a gas at temperature T is simply given by $\sigma_T = (kT/m)^{1/2}$, where k is Boltzmann's constant and m is the mean molecular weight of the observed species.

Most molecular clouds have observed velocity dispersions σ_0 greater than the predicted thermal velocity dispersion σ_T. This excess is referred to as the non-thermal velocity dispersion σ_{NT}, and is given by $\sigma_{NT}^2 = \sigma_0^2 - \sigma_T^2$, since the velocity dispersions add in quadrature. For the smallest cores σ_T dominates, but for cores above a certain radius

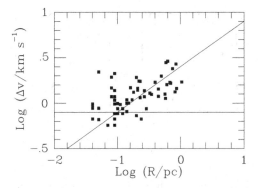

Fig. 4.4. Graph of linewidth versus size on a log-log plot for molecular cloud cores in the Orion region.

(which is typically between 0.01 and 0.1 pc) σ_{NT} is greater than σ_T. Note that none of the other line-broadening mechanisms can have a sufficient effect to cause the observed linewidths.

Figure 4.4 shows a linewidth–size plot for an ensemble of massive molecular cloud cores in the Orion molecular cloud region. The horizontal line on the plot corresponds to the thermal linewidth which would be observed if all the cores were at a temperature of roughly 20 K (a typical average core temperature in a region such as Orion). The sloping line corresponds to $\Delta v \propto R^{0.5}$ (cf. equation 4.45).

Figure 4.4 illustrates how almost all cores of radius greater than about 0.1 pc have linewidths above the horizontal line corresponding to their thermal linewidth. Furthermore, the cores appear to follow the sloping line of equation 4.45. Hence the cores with radius more than 0.1 pc are dominated by the non-thermal velocity dispersion. Thus it is the turbulent motions of molecular clouds which increase with increasing cloud mass and which apparently generate Larson's scaling relations.

We can therefore understand the empirical Larson relations in terms of turbulence and the virial theorem as follows: molecular cloud regions which have greater turbulent internal motions can virially support a greater mass. Hence higher mass cores tend to be seen in these regions (lower mass cores would be preferentially dissipated by the turbulent motions). Lower mass cores on the other hand can more easily survive in regions where the turbulent motions of the interstellar medium are less pronounced. Thus Larson's laws are consistent with a manifestation of the virial theorem in a turbulent medium.

4.6.2 Intermittency

Often spectra require at least a two-Gaussian fit – one for the core of the spectral line shape and one for the wings. Figure 4.5 shows some

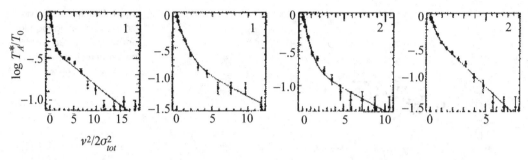

Fig. 4.5. Spectral line profiles of molecular clouds plotted in such a way that a Gaussian line profile would appear as a straight line (the spectra are also shown as two pairs of half-profiles). The fact that none of the spectra can be fitted by a single straight line indicates that there is excess emission in the line wings, which is interpreted as evidence of turbulent motions in the clouds.

typical spectral line profiles, together with a two-Gaussian fit to each spectrum. A single-temperature thermal line spectral profile should be consistent with a single Gaussian. The excess emission in the line wings (i.e. furthest from the ambient velocity of the cloud) indicates that there is a significant amount of high-velocity gas in excess of that which would be predicted by purely thermal motion.

If the wing emission were due to a high-temperature thermal component, the temperatures required would be up to 10^4 K, for which there is no evidence in these molecular clouds. However, one predicted property of turbulent motion in fluids is the presence of temporarily high-velocity material in highly localised regions in time and space. This property is known as the intermittency of the turbulent fluid. The high-velocity spectral line wing emission observed in Figure 4.5 may thus be attributed to the intermittent behaviour of turbulence in molecular clouds.

4.6.3 Turbulent cascades

Turbulent behaviour occurs across a range of size-scales and is predicted to lead to a scale-free geometry. As was seen in Figure 4.2 molecular clouds do show structure on all scales, from the resolution of modern telescopes to the scale of Galactic features such as spiral arms, consistent with the turbulent picture.

One particular form of turbulence involves a scale-free turbulent cascade of eddies in which energy is input on the largest scales (e.g. by supernovae or other mass-losing stars), and cascades down to small scales, where it is dissipated by heating the molecular cloud. The motion of a turbulent cascade in an incompressible fluid produces a characteristic relation between the size-scale of a turbulent eddy L and the velocity dispersion σ of that eddy, of the form

$$\sigma \propto L^{1/3}. \tag{4.52}$$

This is known as the Kolmogorov–Obhukov law.

4.6.4 Fractal structure

Scale-free turbulence predicts the formation of scale-free, or fractal, structure. A fractal is a pattern which repeats on all size-scales, such that it contains no intrinsic scale, and any observation of a fractal should look similar, regardless of the resolution of the observation.

In two dimensions one can define a fractal dimension, $\mathcal{D}$, relating the perimeter, $\mathcal{P}$, of a closed contour in an image, such as that shown in Figure 4.2, with the area, A, enclosed by that contour, such that

$$\mathcal{P} \propto A^{\mathcal{D}/2}. \qquad (4.53)$$

Laboratory experiments on turbulent flows typically show that where turbulence is the dominant process for forming structure within a fluid, then the fractal dimension of that fluid will be $\mathcal{D} \sim 1.36 \pm 0.05$. Likewise, observations of meteorological clouds in the Earth's atmosphere reveal them to have a fractal dimension of $\mathcal{D} \sim 1.35$.

Observations of molecular clouds on size-scales from ≤ 0.1 pc, up to ≥ 100 pc, in various CO isotope transitions, as well as far-infrared dust continuum images, show that they all have a very narrow range of fractal dimension, $\mathcal{D} \sim 1.3$–1.4. This is interpreted as yet more evidence that turbulence is a significant physical process in molecular clouds.

4.6.5 Very small-scale structure

A near-infrared survey of the molecular cloud IC 5146 in Cygnus was carried out (see Figure 4.6), which used a method of counting stars in discrete bins and measuring their colours to determine the extinction A_V in each bin.

Figure 4.6 shows a plot of the dispersion in the measurements of visual extinction A_V versus A_V itself, taken from this survey (do not confuse this dispersion with the velocity dispersions discussed above). It was seen that the 'error' in the measurement of visual extinction A_V was greater than could be accounted for by experimental errors. Furthermore, this 'error' was itself a function of extinction, which increased with increasing A_V. Thus the 'error' is in fact a true dispersion of A_V within each measured bin.

This exact form of the variation of A_V can be explained as being caused by substructure in the extinction on spatial scales smaller than the counting bins. Furthermore, only structure obeying a power law in density in the two directions of the projection of the cloud on the sky would reproduce the exact form of the σ–A_V distribution seen in Figure 4.6. This form of density distribution is what would be predicted by intermittent turbulent behaviour in molecular clouds.

Fig. 4.6. Plot of the
dispersion in extinction
against extinction A_V for the
IC 5146 molecular cloud.

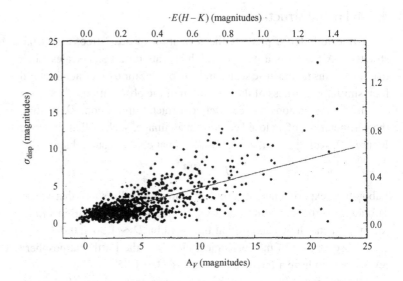

Fig. 4.6. Plot of the dispersion in extinction against extinction A_V for the IC 5146 molecular cloud.

Clearly there are many indicators in observations of molecular clouds which point towards turbulence as one of the dominant physical processes taking place in molecular clouds. There are some difficulties however. Gravity cannot be ignored and this confuses attempts at direct analogies between laboratory studies of turbulence and the astrophysical case. Clear differences exist between the two. For example, normal turbulence is inherently rotational, and an eddy of a particular size has a rotational velocity of the same order as the velocity dispersion. Studies of molecular cloud rotation show that this may not be the case, with observed rotational velocities being less than the typical velocity dispersion. However, it is clear that turbulence does have an important effect on molecular clouds.

4.6.6 Shock fronts

Under certain extreme circumstances in the ISM a shock can occur in the gas. These circumstances include the vicinity of a supernova, or when two turbulent flows of gas collide, for example. A shock occurs when a compression wave advances supersonically into the gas. The compression wave steepens and forms a narrow shock front, a few particle mean-free-paths wide. Gas flows into the shock front supersonically at low density, and out of the shock front subsonically at high density.

There are different types of shocks observed in the ISM. A jump shock, or J-shock, is a shock in which the gas properties change abruptly from one side of the shock to the other. A continuous shock, or C-shock, is one in which the properties vary more gradually from one side to the

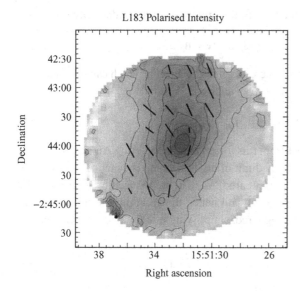

L183 Polarised Intensity

Fig. 4.7. Polarisation map of a molecular cloud core showing the magnetic field direction inferred in the plane of the sky from the measurements.

other. C-shocks tend to occur in regions where the magnetic field plays a significant role in spreading out the effects of the shock over a greater distance.

4.7 Magnetic fields in molecular clouds

We saw in Chapter 1 that there is observational evidence for magnetic fields in the ISM. Figure 4.7 shows a polarisation map of a molecular cloud core in which the polarisation is believed to be due to the magnetic field aligning the dust grains in the core. In this section we will introduce some of the relevant theory of magnetic fields. The branch of physics which deals with magnetic fields in fluids is known as magnetohydrodynamics (MHD). We will discuss how MHD may be relevant to star formation.

We begin with two of Maxwell's equations. Firstly we use Ampère's law

$$\nabla \times \mathbf{B} = \frac{4\pi}{c}\mathbf{J} + \frac{1}{c}\frac{\partial \mathbf{E}}{\partial t} \tag{4.54}$$

where $\mathbf{B}$ is the magnetic flux density, $\mathbf{J}$ is the current density, $\mathbf{E}$ is the electric field strength, and c is the speed of light in a vacuum. Secondly we use Ohm's law, which for a fluid of conductivity σ becomes

$$\mathbf{J} = \sigma \left[\mathbf{E} + \frac{1}{c}(\mathbf{v} \times \mathbf{B}) \right]. \tag{4.55}$$

In this equation the first term on the right-hand side describes the conductivity of a static fluid and the second term incorporates the motion of that fluid at velocity **v**. In the limit of infinite conductivity we have

$$\mathbf{E} + \frac{1}{c}(\mathbf{v} \times \mathbf{B}) = 0. \tag{4.56}$$

In a conducting fluid there are electric currents which can flow and give a force per unit volume $\mathbf{f_m}$ given by

$$\mathbf{f_m} = \frac{1}{c}(\mathbf{J} \times \mathbf{B}). \tag{4.57}$$

We can eliminate **J** from this force equation by using Ampère's law, and remembering that in the limit of high conductivity, such as the ISM, the only electric fields present are those that are induced by motions in the gas with velocity v ($\ll c$), and hence the second term on the right in Ampère's law can be ignored, simplifying it to the magnetostatic case

$$\nabla \times \mathbf{B} = \frac{4\pi}{c}\mathbf{J}, \tag{4.58}$$

which gives us

$$\frac{1}{c}\mathbf{J} = \frac{1}{4\pi}(\nabla \times \mathbf{B}). \tag{4.59}$$

Substituting into the force equation we have

$$\mathbf{f_m} = \frac{1}{4\pi}(\nabla \times \mathbf{B}) \times \mathbf{B} \tag{4.60}$$

or

$$\mathbf{f_m} = -\frac{1}{4\pi}\mathbf{B} \times (\nabla \times \mathbf{B}). \tag{4.61}$$

We can simplify this equation using the vector identity

$$\frac{1}{2}\nabla(\mathbf{B.B}) = (\mathbf{B.}\nabla)\mathbf{B} + \mathbf{B} \times (\nabla \times \mathbf{B}), \tag{4.62}$$

and thus

$$\mathbf{B} \times (\nabla \times \mathbf{B}) = \frac{1}{2}\nabla(\mathbf{B.B}) - (\mathbf{B.}\nabla)\mathbf{B}. \tag{4.63}$$

Then the force equation becomes

$$\mathbf{f_m} = -\nabla\left(\frac{B^2}{8\pi}\right) + \frac{1}{4\pi}(\mathbf{B.}\nabla)\mathbf{B}. \tag{4.64}$$

We can obtain an intuitive physical feel for the meaning of equation 4.64, if we compare it with the Euler equation 4.17. We notice that in the first term on the right-hand side of equation 4.64 the quantity $(B^2/8\pi)$ enters the equation in an identical way to the gas pressure in equation 4.17. Hence this term is known as the magnetic pressure. Any gradient in the magnetic pressure results in a net force on the fluid, just as in the case of the gas pressure. Hence a region of the ISM with a high **B** value tends to be over-pressured relative to neighbouring regions with lower **B**, and will tend to expand.

The second term on the right-hand side of equation 4.64 can be understood as follows. We can write

$$\mathbf{B} = B\mathbf{s}, \tag{4.65}$$

where **s** is a unit vector in the direction of the field. Then we have

$$(\mathbf{B}.\nabla)\mathbf{B} = B\mathbf{s}\frac{d}{dx}(B\mathbf{s})$$

$$= B^2\mathbf{s}\frac{d\mathbf{s}}{dx} + Bs^2\frac{dB}{dx}, \tag{4.66}$$

where x is the direction along the field. For a constant magnetic field, dB/dx is zero and the second term vanishes.

Furthermore, if the field lines are straight, then $d\mathbf{s}/dx$ is zero and the first term vanishes. Hence this term clearly relates to how 'bent' the magnetic field lines are. Furthermore, it is found that the more bent the field lines, the stronger the restoring force. This term is sometimes referred to as the magnetic tension.

The idea that magnetic field lines have a tension rather like tightened strings on a bow leads to the idea that tranverse waves could travel along the field lines like waves on a string. These waves are known as Alfvén waves. Such tranverse waves can have two orthogonal polarisation states, or in general a sum of both at random phase and hence 'torsional' waves. Alfvén waves are believed to play an important role in the turbulent support of the ISM, and torsional Alfvén waves play an important role in star formation by carrying away excess angular momentum, and hence allowing a cloud core to collapse.

In a highly conducting medium such as the ISM we can assume 'flux freezing' – that is, that the magnetic field and the gas move together as a single magnetised fluid, or plasma. Hence, if we take a unit volume of that fluid and compress it orthogonal to the field, the magnetic field strength increases – i.e. $B \propto 1/\text{volume}$. Hence

$$B \propto \rho. \tag{4.67}$$

Note that this assumes a compressible medium. Thus if we have an initial density ρ_0 and field strength B_0, we have

$$\frac{B}{B_0} = \frac{\rho}{\rho_0}. \tag{4.68}$$

Thinking of the ISM as a single fluid plasma allows us to also consider compression waves travelling through the ISM, known as magnetosonic waves by analogy with sound waves in a non-magnetised fluid.

We note that the sound velocity a_0 in a compressible fluid is given by

$$a_0^2 = \frac{dP}{d\rho}. \tag{4.69}$$

If we assume that the magnetic pressure dominates and we remember that the magnetic pressure P_m is given by

$$P_m = \frac{B^2}{8\pi} = \frac{B_0^2 \rho^2}{8\pi \rho_0^2}, \tag{4.70}$$

we find that the wave velocity v_A is given by

$$v_A^2 = \frac{d}{d\rho}\left(\frac{B_0^2 \rho^2}{8\pi \rho_0^2}\right) = \frac{B_0^2 \rho}{4\pi \rho_0^2}. \tag{4.71}$$

Thus in the case of $B = B_0$, as we have here, we can put $\rho = \rho_0$, and we have

$$v_A = \frac{B_0}{(4\pi \rho_0)^{1/2}}. \tag{4.72}$$

This is known as the Alfvén wave velocity.

The virial theorem (equation 4.37) in the presence of a magnetic field requires an extra term for the magnetic energy, E_m, so that

$$\Omega_G + 2\mathcal{K} + E_m = 0. \tag{4.73}$$

By analogy with thermodynamics, in which the quantity PV has the dimensions of energy, we deduce that the magnetic energy E_m is given by

$$E_m = P_m V = \frac{B^2 V}{8\pi}. \tag{4.74}$$

We can use this to estimate the stability of a molecular cloud core. Note that we are ignoring surface terms for simplicity.

Consider a uniform density, spherically symmetric cloud of mass M, initial radius R_0 and volume V_0, threaded by a magnetic field of strength B_0, and consider a slight perturbation to compress the sphere to a radius R. If the magnetic pressure is the dominant support mechanism (i.e. $E_m \gg \mathcal{K}$), then for equilibrium, the magnetic virial theorem reduces to

$$E_m \simeq -\Omega_G. \tag{4.75}$$

For a uniform sphere the gravitational potential energy is given by

$$\Omega_G \simeq -\frac{3GM^2}{5R}. \tag{4.76}$$

A slight compression forces the field lines closer together, and hence increases the magnetic field strength. From consideration of conservation of magnetic flux we know that the total flux threading a given area is a conserved quantity. Hence

$$B_0 R_0^2 = B_R R^2, \tag{4.77}$$

where B_R is the new magnetic field strength after the sphere has been compressed slightly to radius R. Thus

$$B_R = B_0 \left(\frac{R_0}{R} \right)^2. \tag{4.78}$$

Hence, within R, the magnetic energy is given by

$$E_m = \frac{B^2 V}{8\pi} = \frac{1}{8\pi} B_0^2 \left(\frac{R_0}{R} \right)^4 \frac{4}{3} \pi R^3 = \frac{B_0^2 R_0^4}{6R}. \tag{4.79}$$

In actual fact there is an extra term of the same magnitude for the magnetic field strength between R and R_0 such that the magnetic energy is

$$E_m = \frac{B_0^2 R_0^4}{3R}. \tag{4.80}$$

So for equilibrium we require that

$$\frac{B_0^2 R_0^4}{3R} = \frac{3G M_c^2}{5R}. \tag{4.81}$$

Therefore

$$M_c = \left(\frac{5 B_0^2 R_0^4}{9G} \right)^{1/2} = B_0 R_0^2 \left(\frac{5}{9G} \right)^{1/2}. \tag{4.82}$$

The mass M_c is known as the magnetic critical mass. We can also define a critical mass-to-flux ratio in terms of

$$\frac{M_c}{B_0 R_0^2} = \left(\frac{5}{9G} \right)^{1/2}. \tag{4.83}$$

The significance of M_c is that if the mass of the cloud is less than M_c then the cloud is stable against collapse. However, if the magnetic field is the dominant mechanism of support, and the cloud mass is greater than M_c, then it is unstable to collapse. In the next chapter we will study the manner in which this collapse proceeds.

4.8 Chemistry in molecular clouds

4.8.1 Gas-phase chemistry

We have already mentioned that molecular clouds contain molecules other than H_2. The most abundant of these in the gas phase is carbon monoxide, CO. We know about this because of the molecular rotational transitions that we detect in the millimetre waveband and the molecular vibrational transitions that we detect in the infrared.

However, for many years it was a mystery as to how molecules could form in the relatively low-density environments of the interstellar medium in molecular clouds. The problem is that when two atoms collide in the gas phase the most likely outcome is that they will simply bounce off. A third body is normally required to carry away the excess energy.

The most common atom is hydrogen, and hence the most common molecule is H_2. The majority of molecular hydrogen in fact forms on dust grains. The grain acts as a sink for the excess energy. Molecules can then be released from the grains back into the gas phase. Molecular hydrogen is formed on the surface of dust grains by means of the reaction

$$H + H \rightarrow H_2. \tag{4.84}$$

We will deal further with grain surface chemistry in the next section, but first we deal with gas-phase reactions.

A way of increasing the collisional cross-section in the gas phase is if one of the participants in the reaction is ionised. The source of the ionisation may occur by means of UV ionisation by the interstellar radiation field, or perhaps more importantly in molecular cloud interiors, ionisation by cosmic rays (see section 1.3), such as

$$H_2 + cr \rightarrow H_2^+ + e^- + cr, \tag{4.85}$$

where we have denoted the cosmic ray as cr. The ionisation helps to drive the subsequent chemistry. For example

$$H_2^+ + H_2 \rightarrow H_3^+ + H, \tag{4.86}$$

which produces the highly reactive molecular ion H_3^+. This can drive the oxygen chemistry, leading to the formation of the water molecule, H_2O, and the OH radical by the following route

$$H_3^+ + O \rightarrow OH^+ + H_2. \tag{4.87}$$

An alternative formation mechanism for OH^+ could be by means of ionised oxygen, using

$$H_2 + O^+ \rightarrow OH^+ + H. \tag{4.88}$$

Then

$$OH^+ + H_2 \rightarrow H_2O^+ + H, \tag{4.89}$$

followed by

$$H_2O^+ + H_2 \rightarrow H_3O^+ + H, \tag{4.90}$$

and

$$H_3O^+ + e^- \rightarrow OH + 2H, \tag{4.91}$$

or

$$H_3O^+ + e^- \rightarrow H_2O + H. \tag{4.92}$$

The carbon chemistry follows a similar route. For example

$$C^+ + H_2 \rightarrow CH_2^+ + h\nu, \tag{4.93}$$

where we have denoted an emitted photon by $h\nu$, and then

$$CH_2^+ + H_2 \rightarrow CH_3^+ + H, \tag{4.94}$$

and

$$CH_3^+ + e^- \rightarrow CH + 2H, \tag{4.95}$$

or

$$CH_3^+ + e^- \rightarrow CH_2 + H. \tag{4.96}$$

The major pathways to forming CO appear to be

$$C + OH \rightarrow CO + H, \tag{4.97}$$

or

$$C^+ + OH \rightarrow CO^+ + H. \tag{4.98}$$

These are known as exchange reactions. The latter proceeds to CO via

$$CO^+ + H_2 \rightarrow HCO^+ + H, \tag{4.99}$$

and

$$HCO^+ + e^- \rightarrow CO + H, \tag{4.100}$$

although in fact HCO^+ can be the dominant ion in molecular clouds.

Other neutral exchange reactions are important for sulphur chemistry, such as

$$OH + S \rightarrow SO + H, \tag{4.101}$$

and also for nitrogen chemistry

$$CH + N \rightarrow CN + H. \tag{4.102}$$

This can then go on to form N_2 by means of

$$CN + N \rightarrow N_2 + C. \tag{4.103}$$

Fig. 4.8. Column density profile across a molecular cloud core, computed in two different ways. The crosses represent measurements made in dust continuum observations, and the circles are measurements of CO. Note how the CO profile lies below the dust continuum profile, implying that there is much less CO (by a factor of 2–3) in the gas phase at the centre of the core than would be predicted. This is believed to be due to the CO freezing out in solid form onto the surfaces of dust grains in the densest part of this core.

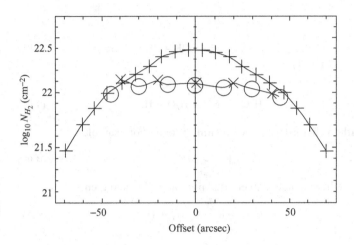

An alternative route is

$$N + OH \rightarrow NO + H, \tag{4.104}$$

followed by

$$NO + N \rightarrow N_2 + O. \tag{4.105}$$

All of these different types of reaction are believed to take place in the gaseous interstellar medium.

4.8.2 Grain surface chemistry

However, astronomers have now observed more than 100 different molecular species in the ISM, requiring far more complex reaction chains than the simple reactions outlined above. Some of the larger molecules have abundances much higher than can be explained by gas-phase reactions alone.

The problem was solved when it was realised that molecules could form on the surfaces of interstellar dust grains, with the grain acting as a heat sink for the excess energy. Grain surface reactions are thus important for interstellar chemistry. Figure 4.8 shows an example of some of the evidence we have for molecules being depleted in the gas phase, and hence by inference existing in solid phase on the surfaces of dust grains. Figure 4.8 shows two profiles across a molecular cloud core, one computed from continuum measurements of the dust and the other from spectral line measurements of the gas, in this case CO. The CO profile lies below the dust profile in the centre of the core, hence implying a reduction in gas-phase CO as it depletes onto dust grains.

The main physical processes involved in grain surface chemistry are still a subject of intense research, but they can be divided into four main types: accretion of the atom or molecule onto the dust grain surface; movement of the atom or molecule across the surface; reaction between atoms and/or molecules; and the return of the new molecule to the gas phase.

If an atom hits a dust grain and sticks it is said to be 'adsorbed' onto the grain surface. If it strikes a site where it can bind, the process is known as 'chemisorption'. If it is held by van der Waals forces the process is known as 'physisorption'. The latter kind of binding allows an atom to move across the surface easily and interact with another atom either by 'hopping' or quantum tunnelling. This is known as 'diffusion' and is dominated by hydrogen chemistry, hydrogen being the lightest and most mobile atom. This turns O into H_2O, C to CH_4, N to NH_3 and S to H_2S.

CO formed in the gas phase is also adsorbed onto grain surfaces, and the chief chemical components observed on grain surfaces are in fact CO and H_2O. The chemicals built up on grain surfaces are known as grain mantles, and the thickness of water and CO ice mantles can be such as to more than double the size of the original grain. Whilst CO is more common in the gas phase, it appears that CO_2 is more common on grain surfaces, where it is presumably formed.

Atoms or molecules which are strongly bound, or chemisorbed, are not nearly so mobile as those which are physisorbed, and they can only interact with other atoms or molecules which happen to land very close by. These processes are therefore slower and less efficient.

Finally, the removal of molecules from grain surfaces and back into the gaseous phase is known as 'desorption'. At high densities UV induced desorption is inefficient due to the high extinction. However, as soon as a star forms and heats its surroundings then the process known as 'thermal desorption' becomes significant. Temperatures of ~100 K or more are required to return significant grain mantle material to the gaseous ISM.

Other desorption processes include the action of cosmic rays or UV photons from the interstellar radiation field to remove molecules from grain surfaces. These provide a background, continuous desorption, and the latter is most significant on the edges of molecular clouds, or near newly formed high-mass stars. Heating caused by molecule formation on dust grains can also lead to desorption.

The destruction of molecules by UV photons in the general interstellar radiation field is the main mechanism responsible for returning molecular material to the atomic state. Two processes reduce the effects of this in molecular cloud interiors. One is the process of 'self-shielding', whereby molecules at the edges of a cloud absorb the photons with high enough energy to dissociate molecules, thereby shielding the molecules in the cloud interiors. The other is absorption of high-energy photons by dust grains, the so-called 'extinction' which renders the centres of molecular clouds 'dark'.

4.8.3 Carbon chemistry

Particular mention should be made of carbon chemistry in the ISM. It is of interest as the basis of all life as we know it, although no such life elsewhere has yet been found. Nonetheless the ISM appears to contain a rich carbon chemistry.

Carbon is observed in many forms in the ISM. The first to be discovered was CN, followed by CO, and then long chain compounds such as the alcohols CH_3OH, C_2H_5OH, etc., of the form $C_nH_{2n+1}OH$. These are the common molecules methanol and ethanol and so on. Astronomers sometimes amuse themselves by calculating the amount of naturally occurring alcohol in a molecular cloud. The answer can be quite a significant quantity.

Another important form of carbon in the ISM is in molecules derived from multiple benzene rings of C_6 and their various hydrides – the so-called 'aromatic' compounds. These can grow to very large sizes, with molecules containing tens or even hundreds of atoms being observed. These multiple C_6 rings can form large sheet-like molecules known as polycyclic aromatic hydrocarbons, or PAHs for short.

One particular variant of a PAH occurs when a sheet-like molecule folds over into a spherical shape, forming C_{60}, known as a Buckminster fullerene. There are many types of Buckminster fullerene, depending on the level of hydrogenation, up to $C_{60}H_{60}$. These are sometimes known as buckyballs, and some astronomers believe that up to $\sim 1\%$ of all interstellar carbon may be in this form.

Another major form of carbon believed by many astronomers to exist in the ISM is graphite. The so-called extinction curve which plots interstellar extinction as a function of wavelength has a peak at around 220 nm, which matches laboratory measurements of graphite. This leads some astronomers to think that not all dust grains are silicates, and some may be graphitic. This carbon may alternatively exist in amorphous form on the surfaces of dust grains. In this form it is known as hydrogenated amorphous carbon (HAC).

However, there have been claims that this spectral feature can also be matched by a simple PAH, C_{24}, known as coronene. The hydrides of C_{24} may also be responsible for other interstellar spectral features known as diffuse interstellar bands (DIBs). These features have been observed for many years in the ISM, but the exact molecules responsible for them remain a matter of debate. For example, it is highly likely that much of the carbon in the interstellar medium could be amorphous, more like soot. The debate over the nature of the interstellar carbon will no doubt continue for some years to come.

4.8.4 Chemistry and star formation

The study of interstellar chemistry could occupy a whole textbook in its own right, so we refer the reader to the list of further reading at the end of the chapter for more information. However, we conclude this section by noting that chemistry impinges directly upon the star-formation process in a number of different ways.

Chemistry affects the ionisation level of a molecular cloud. This in turn affects the manner in which the matter couples to the magnetic field, which may be responsible for retarding the collapse that forms a star.

Chemistry affects the optical properties of dust grains in the ISM by coating the grains with mantles containing water ice, CO ice and many other molecules. This must be accounted for when using measurements of the millimetre and submillimetre continuum to calculate the mass of a molecular cloud.

Chemistry helps the study of star formation by providing many different molecules whose spectra can be studied. Remember that it is very difficult to observe molecular hydrogen, H_2, directly, due to it not having a permanent dipole moment. Therefore most molecular clouds are studied in transitions of CO or other molecules.

Chemistry affects the micro-physics of a molecular cloud. In particular, it alters the equation of state of a cloud in complex ways that are still not fully understood. Most importantly for star formation, this affects the ability of a cloud to cool and radiate away its internal energy, allowing collapse and star formation to proceed.

Finally, it is the carbon chemistry discussed in the previous section that provides the building blocks for life itself. We do not know how this occurred on Earth, but without it there would be no astronomers to study star formation.

Recommended further reading

We recommend the following texts to the student for further reading on the topics presented in this chapter.

Falgarone, E., Boulanger, F. and Duvert, G. (1990). *Fragmentation of Molecular Clouds and Star Formation*. International Astronomical Union Symposium vol. 147. Dordrecht: Kluwer.

James, R. A. and Millar, T. J. (1991). *Molecular Clouds*. Cambridge: Cambridge University Press.

Kahn, F. D. (1985). *Cosmical Gas Dynamics*. Utrecht: VNU Science.

Landau, L. D. and Lifshitz, E. M. (1987). *Fluid Mechanics*, 2nd edn. Amsterdam: Elsevier Butterworth–Heinemann.

Mandelbrot, B. B. (1982). *The Fractal Geometry of Nature*. San Francisco: W. H. Freeman.

Mestel, L. (2003). *Stellar Magnetism*. International Monographs on Physics, vol. 99. Oxford: Oxford University Press.

Shaw, A. M. (2006). *Astrochemistry from Astronomy to Astrobiology*. Chichester: Wiley.

Shore, S. N. (2007). *Astrophysical Hydrodynamics: An Introduction*. Weinheim: Wiley-VCH.

Chapter 5

Fragmentation and collapse – the road to star formation

5.1 The road to star formation

Thus far we have studied the places where stars form – molecular clouds. We have discussed the ways in which molecular clouds can be observed. We have explored the various constituents of molecular clouds – gas, dust, magnetic fields, cosmic rays and electromagnetic radiation. We have, so to speak, assembled the ingredients. In this chapter we discuss how these ingredients might come together to begin to form a star.

In the first half of the chapter we discuss theoretical considerations. We consider the collapse of an isothermal sphere of gas, ignoring the effects of rotation and magnetic fields, and we examine qualitatively what happens. We describe the method of solving the problem using similarity solutions.

We go on to discuss hierarchical fragmentation, as a means of breaking a large molecular cloud into an ensemble of stars. We also discuss the thermodynamics of protostellar gas, and explain how the minimum mass for star formation might be determined by the protostellar gas becoming optically thick to its own cooling radiation. We discuss the manner of the collapse to form a star and the possible effects of a magnetic field on this process.

In the second half of the chapter we examine some of the observational evidence. At the end we consider the initial mass function for stars. Note that in this chapter we concentrate mainly on relatively low-mass stars, i.e. stars of less than a few times the mass of the Sun. In Chapter 6 we continue to discuss relatively low-mass star formation.

Then in Chapter 7 we turn to the questions associated with high-mass star formation.

5.2 Theoretical collapse solutions

We begin by considering an isothermal, uniform-density, non-rotating, non-magnetised, spherically symmetric cloud. If we also assume that the cloud is initially pressureless, then it will collapse in a free-fall time, t_{ff}, as we saw in the previous chapter (equation 4.43), given by

$$t_{ff} = \left(\frac{3\pi}{32G\rho_0} \right)^{1/2} \tag{5.1}$$

where ρ_0 is the initial density. The equation of motion for material within the cloud is given by

$$\frac{d^2r}{dt^2} = -\frac{GM}{r^2} \tag{5.2}$$

at any radius r. Recall also from the previous chapter (equation 4.76) that the gravitational potential energy, Ω_G, of a sphere of mass M and radius R is

$$\Omega_G \simeq -\frac{3GM^2}{5R}. \tag{5.3}$$

At any given radius the continuity equation (equation 4.9) is

$$\frac{\partial \rho}{\partial t} + \rho \nabla . \mathbf{v} + \mathbf{v} . \nabla \rho = 0, \tag{5.4}$$

giving us the density at that point as a function of time.

Once the sphere starts to collapse we can no longer ignore the pressure. The motion of the fluid as a whole is then given by Euler's equation (equation 4.18), which takes the form

$$\frac{\partial \mathbf{v}}{\partial t} + (\mathbf{v} . \nabla)\mathbf{v} + \frac{\nabla P}{\rho} - \mathbf{g} = 0, \tag{5.5}$$

where $\mathbf{g}$ is the gravitational acceleration experienced by a parcel of gas. These equations can be solved numerically to follow the collapse of a theoretical cloud.

One of the key considerations in all such theoretical calculations is knowing how to deal with the boundary conditions. One solution is to hold the edge of the cloud fixed in space and follow the evolution. We can obtain a qualitative understanding of what might happen. As the material falls towards the centre, the density at the edge of the cloud (which is held fixed) drops, while the density nearer to the centre of the cloud rises.

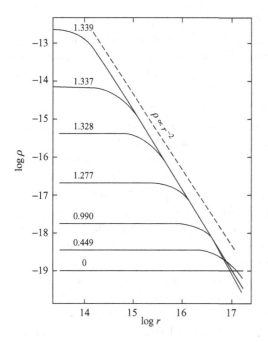

Fig. 5.1. The logarithmic variation of density with radius as a function of time in one model of a collapsing isothermal cloud, in which the outer boundary is held fixed. Each curve is labelled with the time elapsed since collapse began (in arbitrary units). Note how the density profile approximates to r^{-2} in the outer parts.

This sets up a pressure gradient in the outer parts of the cloud. This in turn slows the collapse in the outer parts with respect to the free-fall solution. The density therefore rises more rapidly still in the inner parts and the pressure gradient continues to retard the collapse further out. The collapse in the centre continues in a roughly free-fall manner.

We can see from equation 5.1 that the free-fall time has an inverse dependence on density. Hence the higher-density regions collapse more quickly and the density distribution becomes more centrally peaked. In fact the density distribution follows the form shown in Figure 5.1. It can be seen that in this case the density profile approximates to the form

$$\rho \propto r^{-2} \tag{5.6}$$

in the outer parts of the cloud. This is one of a family of solutions that are known as similarity solutions or self-similar solutions. The name arises because the solution appears similar on different size-scales. Note that in the model in Figure 5.1, at late times the r^{-2} dependence extends over three orders of magnitude in radius. Hence, it is said to be self-similar across these size-scales. The similarity solution breaks down at the centre, since $\rho \to \infty$ as $r \to 0$, which is clearly non-physical. Hence this is also sometimes known as the singularity solution, and equation 5.6 is sometimes known as the equation of a singular isothermal sphere, or SIS.

However, the model breaks down before this occurs for another reason. The collapse generates heat due to the compression of the gas (sometimes known as P-V work). This heat can be radiated away as long as the cloud remains optically thin. But at sufficiently high density the cloud becomes optically thick, and it can no longer radiate away the excess heat. Hence the densest parts of the cloud heat up and the assumption of isothermality breaks down.

When the central part of the cloud becomes optically thick it rapidly heats up. The internal energy of the central region increases until it balances the sum of the gravitational potential and the infall kinetic energy of the central region (this is similar to the virial condition – see equation 4.37). At this point collapse at the centre of the cloud stops.

The optically thick, non-collapsing region at the centre is sometimes referred to as the 'first core'. Material from the rest of the infalling cloud continues to fall onto the surface of the first core, where it is brought to an abrupt halt at a shock front. The heat generated at this shock is radiated away in the infrared.

The predicted subsequent evolution of the collapsing cloud is highly model-dependent. In fact, the model discussed above is so highly idealised that it is unlikely to resemble any real clouds. For the same reason, similarity solutions on the whole are no longer generally used to model collapsing clouds. They have largely been replaced by more sophisticated computer models. We return to the topic of cloud collapse and the subsequent evolution in Chapter 6.

5.3 The minimum mass of a star

5.3.1 Hierarchical fragmentation

In Chapter 4 we discussed the stability of a cloud against collapse in terms of the Jeans criterion and the virial theorem. Given a uniform three-dimensional background medium, having density ρ_0 and isothermal sound speed a_0, the Jeans radius, R_J, given by

$$R_J \simeq \left(\frac{15}{4 \pi G \rho_0} \right)^{1/2} a_0, \tag{5.7}$$

and the Jeans mass, M_J, given by

$$M_J \simeq \left(\frac{375}{4 \pi} \right)^{1/2} \frac{a_0^3}{G^{3/2} \rho_0^{1/2}}, \tag{5.8}$$

define the smallest lump which can collapse (i.e. the smallest lump for which it is *thermodynamically viable* to collapse). Collapse will actually occur only if there are no other competing processes with shorter time-scales. Apart from purely numerical factors, these expressions are the same as those we derived in Chapter 4, where we considered the Jeans instability in one dimension.

Suppose now that an interstellar cloud becomes gravitationally unstable ($M_0 > M_J$) and starts to contract, but remains isothermal (and therefore a_0 is constant). Because the density in the cloud increases, the Jeans mass decreases, since $M_J \propto \rho_0^{-1/2}$. Consequently, parts of the original cloud become gravitationally unstable in their own right. Therefore they can start to contract on themselves, causing the original cloud to fragment. Moreover, the process can repeat itself again and again, as long as the gas remains isothermal – see Figure 5.2.

The gas remains isothermal as long as it can radiate away energy as fast as it is being released by gravitational contraction. Once the fragments become sufficiently opaque, their cooling radiation becomes trapped, and they heat up until they are close to hydrostatic equilibrium; that is to say that the mass of an individual fragment, M_{frag}, becomes comparable with its Jeans mass, M_J. Fragmentation then ceases. We now derive mathematically the density at which this occurs.

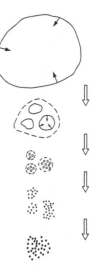

Fig. 5.2. Hierarchical fragmentation breaks up a large cloud into smaller fragments.

5.3.2 Contraction of a marginally unstable fragment

We consider a uniform density, spherically symmetric, cloud fragment, which is initially in equilibrium and surrounded by a medium with the same uniform density. To test the fragment's stability against collapse we consider what happens when it experiences a slight perturbation. We introduce $\phi(R)$, which is a function controlling radial excursions of the cloud fragment away from equilibrium and is a function of radius R from the fragment centre.[†]

Consider a marginally Jeans-unstable fragment with mass $M_0 = M_J$ and initial radius $R_0 = R_J$, starting to collapse. In this case the function $\phi(R)$ can be shown to have the form

$$\frac{\phi(R)}{a_0^2} = -\left[\left(\frac{R}{R_0}\right)^{-1} - 1\right] - \ln\left(\frac{R}{R_0}\right) + \frac{1}{3}\left[\left(\frac{R}{R_0}\right)^3 - 1\right]. \quad (5.9)$$

We have

$$\ddot{R} = -\frac{d\phi}{dR}, \quad (5.10)$$

and hence

$$2\dot{R}\ddot{R} = -2\frac{d\phi}{dR}\dot{R}. \quad (5.11)$$

[†] Do not confuse $\phi(R)$ with the IMF, $\phi(M)$, or the profile function, $\phi(v - v_0)$.

This can be integrated to give

$$\dot{R}^2 = 2 \left[\phi(R_0) - \phi(R)\right] = -2\,\phi(R), \qquad (5.12)$$

since by definition $\phi(R_0) = 0$. Substituting for $\phi(R)$ from equation 5.9, and putting $R_0 = R_J$, we have

$$\dot{R}^2 = 2a_0^2 \left\{ \left[\left(\frac{R}{R_J}\right)^{-1} - 1\right] + \ln\left(\frac{R}{R_J}\right) - \frac{1}{3}\left[\left(\frac{R}{R_J}\right)^3 - 1\right]\right\}. \qquad (5.13)$$

Then, introducing the compression factor, f, where

$$f = \frac{R_J}{R}, \qquad (5.14)$$

we obtain

$$\dot{R} = g(f)\,a_0, \qquad (5.15)$$

where $g(f)$ is simply a function given by

$$g(f) = 2^{1/2} \left\{[f - 1] - \ln[f] + \frac{1}{3}\left[1 - f^{-3}\right]\right\}^{1/2}. \qquad (5.16)$$

Thus we have a relation between the rate of compression, $\dot{R}$, and the compression factor, f, in terms of the Jeans radius, R_J, and the isothermal sound speed, a_0.

5.3.3 The compressional heating rate

For a spherical cloud with mass M_0 and isothermal sound speed a_0, the compressional heating rate, $\mathcal{H}_{\mathrm{comp}}$, is

$$\mathcal{H}_{\mathrm{comp}} = -P\frac{dV}{dt} = -\frac{3\,M_0\,a_0^2}{4\pi\,R^3}\,4\pi\,R^2\,\frac{dR}{dt} = -3\,M_0 a_0^2\,\frac{\dot{R}}{R}. \qquad (5.17)$$

Substituting for R and $\dot{R}$ from equations 5.14 and 5.15, and putting $M_0 = M_J$, we obtain (after a little algebra)

$$\mathcal{H}_{\mathrm{comp}} = f g(f)\,\frac{15\,a_0^5}{G}. \qquad (5.18)$$

5.3.4 Radiative cooling rate

The fragment cannot radiate more efficiently than a blackbody at the same temperature. Therefore its luminosity satisfies

$$\mathcal{L} \stackrel{<}{\sim} \mathcal{L}_{\mathrm{max}} = 4\pi\,R^2\,\sigma_{\mathrm{SB}}\,T^4, \qquad (5.19)$$

where σ_{SB} is the Stefan–Boltzmann constant, given by

$$\sigma_{\mathrm{SB}} = \frac{2\pi^5\,k^4}{15\,c^2\,h^3}. \qquad (5.20)$$

In reality the optical depth is of order unity and so we can put $\mathcal{L} = \mathcal{L}_{\text{max}}$. Then if we put $R = R_{\text{J}}/f$, equation 5.19 becomes

$$\mathcal{L} = \mathcal{L}_{\text{max}} = f^{-2} \frac{15\, a_0^2\, \sigma_{\text{SB}}\, T^4}{G\, \rho_0}. \tag{5.21}$$

5.3.5 Condition for isothermality to be maintained

Isothermality can only be maintained in a fragment if the compressional heating rate is less than, or on the order of, the radiative cooling rate

$$\mathcal{H}_{\text{comp}} \overset{<}{\sim} \mathcal{L}_{\text{max}}. \tag{5.22}$$

Substituting from equations 5.18 and 5.21, we obtain a constraint on the density in the background in which the fragment is trying to collapse

$$\rho_0 \overset{<}{\sim} \rho_{\text{max}} = \frac{\sigma_{\text{SB}}\, T^4}{f^3\, g(f)\, a_0^3}. \tag{5.23}$$

Above the density ρ_{max} the fragment can no longer radiate away its compressional luminosity and so it must heat up.

5.3.6 The minimum mass

This upper limit on the density translates into a lower limit on the Jeans mass, M_{J}

$$M_{\text{J}} > M_{\text{min}} = \left[f^3 g(f) \right]^{1/2} \frac{75}{\pi^3}\, c \left(\frac{h}{2G} \right)^{3/2} \left(\frac{kT}{\bar{m}^9} \right)^{1/4}, \tag{5.24}$$

where we have substituted σ_{SB} from equation 5.20.

We consider compression factors f in the range 2–4, since this translates into an increase in ρ by between 8 and 64, and hence a reduction in M_{J} by between ~3 and 8. In other words, we assume that at each level of the hierarchy, an individual fragment spawns between three and eight subfragments. Then

$$3 \overset{<}{\sim} \left[f^3 g(f) \right]^{1/2} \overset{<}{\sim} 11, \tag{5.25}$$

and we adopt

$$\left[f^3 g(f) \right]^{1/2} \sim 4 \tag{5.26}$$

as a representative value. Equation 5.24 then becomes

$$M_{\text{frag}} > M_{\text{min}} \simeq 10\, c \left(\frac{h}{2G} \right)^{3/2} \left(\frac{kT}{\bar{m}^9} \right)^{1/4}. \tag{5.27}$$

For contemporary, Population I star-forming gas, $T \sim 10\,\text{K}$ and $\bar{m} \sim 4 \times 10^{-27}\,\text{kg}$, so $M_{\text{min}} \simeq 0.015 M_{\odot}$. We note that this is less than the hydrogen-burning limit for a star, which is approximately $0.08M_\odot$. Stars with mass below this limit are known as brown dwarf stars (see Chapter 8). So this theoretical calculation predicts the existence of brown dwarf stars.

5.4 Effects of the magnetic field

In Section 4.7 we discussed the presence of magnetic fields in molecular clouds. There we assumed that the magnetic field is frozen into the gas. There is evidence in some regions that this is in fact the case. Remember that even ionised material can flow along the field lines relatively freely. In theory this would tend to produce a collapse which proceeds preferentially parallel to the magnetic field, with there being more resistance to collapse perpendicular to the field.

Hence collapsing clouds should roughly resemble flattened oblate spheroids, with their long axes perpendicular to the magnetic field. If these oblate spheroids were then to begin to contract perpendicular to their field directions, dragging the field with them, then this would lead to a pinching, or narrowing of the field lines at the centre. Further out along the magnetic field, no such pinching would be seen.

This leads to a shape of the magnetic field lines known as the 'hour-glass', or 'egg-timer' shape. In some regions that is exactly what is seen. Figure 5.3 shows a polarisation map tracing the magnetic field in the star-forming region NGC 1333. In this map one can clearly discern the hour-glass shape of the magnetic field, indicating that collapse has happened first along the field lines, and subsequently perpendicular to the field lines at the centre. This is exactly as predicted from magnetic flux freezing into the gas.

5.4.1 Ion-neutral drift

When the fraction of ionised particles is very low, the assumption of magnetic flux freezing can break down. The reason is that in a real gas there is a mix of ions and neutral molecules. The ions are influenced by the magnetic field, and gyrate around the lines of force with the cyclotron frequency. Since they are free to move along the field lines, their general motion is roughly helical.

However, the neutrals are free to cut across field lines and drift according to pressure gradients or gravitational acceleration. Thus differential motion between ions and molecules is possible, and is called ambipolar diffusion. In most environments the two components are closely coupled through collisions, so that any motion of the ionic component is quickly transferred to the neutral molecules through collisions,

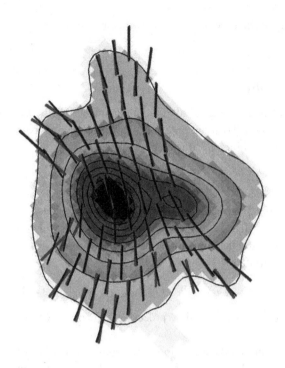

Fig. 5.3. Map of the magnetic field in the star-forming region NGC 1333. The magnetic field direction is implied from polarisation measurements. The two sets of vectors are simply measurements at two different wavelengths. Note the hour-glass, or egg-timer, shape, indicating that some collapse has occurred perpendicular to the field direction, causing the field to be pinched in at the centre. This is a characteristic indicator of magnetic flux freezing, where the field and the matter move together.

and vice versa, and negligible diffusion occurs. But in some cases the coupling breaks down and the ionic and neutral components can drift past each other.

The effects of this can be understood with a simple model. Consider a quasi-equilibrium configuration where a magnetic field is supporting a gas in a gravitational field. The ions are tied to the field lines and gyrate around them. The neutrals, however, are free to accelerate in the direction of the gravitational force. The velocity reached by a neutral (before it collides with an ion and has its velocity randomised) is given by

$$v \sim gt \qquad (5.28)$$

where v is the neutral's velocity, g is the gravitational acceleration and t is the time between collisions. Under the assumption that the thermal velocities greatly exceed the drift velocity, the collision time can be estimated by averaging the cross-section (which is in general a function of the collision energy) and the thermal velocity distribution. Then

$$t = \frac{1}{n_i \langle \sigma u \rangle} \qquad (5.29)$$

where n_i is the number density of ions, and $\langle \sigma u \rangle$ is the average over a Maxwellian velocity distribution of the product of the collision cross-section and the thermal velocity. These two equations define the drift velocity.

5.4.2 Ambipolar diffusion

To put this into context, consider a cloud of radius R and gas density n_H (mainly neutral H and H_2). Then the time for the neutrals to drift past the ions is given by the ambipolar diffusion time-scale

$$\tau_{AD} \sim \frac{R}{v} = \frac{R n_i \langle \sigma u \rangle}{(GM/R^2)} = \frac{3 \langle \sigma u \rangle}{4 \pi G m_H} x, \qquad (5.30)$$

where $x = n_i/n_H$ is the fractional ionisation and m_H is the mass of the hydrogen atom. Note that the diffusion time-scale depends only on the fractional ionisation and the temperature of the gas (through $\langle \sigma u \rangle$). Inserting values appropriate for $T = 10\,K$ yields

$$\tau_{AD} = 7.3 \times 10^{13}\,\text{yr}. \, x \qquad (5.31)$$

Thus ambipolar diffusion is a very slow process except in regions of very low ionisation.

However, it turns out that the ionisation expected in dense cores of molecular clouds is indeed very low. In dark interstellar clouds from which the background UV interstellar radiation is excluded (because of high extinction), ions are produced mainly by cosmic ray (CR) ionisation of H_2 via the reaction

$$H_2 + cr \rightarrow H_2^+ + e^- + cr. \qquad (5.32)$$

This is followed rapidly by the formation of H_3^+, by means of

$$H_2^+ + H_2 \rightarrow H_3^+ + H \qquad (5.33)$$

followed by protonation reactions which yield carbon and oxygen bearing molecules, such as HCO^+. Dissociative recombinations with electrons then produce neutral species such as CH_2 and H_2O. The rate of production of ions per molecule by cosmic rays is independent of the gas density, but the recombination rate depends on the density of electrons. Making the simplifying assumption that only one type of ion is present, the recombination rate per unit volume is given by

$$R_R = 8.2 \times 10^{-7}\,\text{m}^3\,\text{s}^{-1}\,n_i n_e = 8.2 \times 10^{-7}\,\text{m}^3\,\text{s}^{-1}\,x^2 n_{H_2}^2 \qquad (5.34)$$

where n_i, n_e and n_{H_2} are the volume number densities of ions, electrons and H_2 molecules respectively, and the rate cofficient is that for HCO^+.

The rate of cosmic ray ionisation per unit volume is estimated to be

$$I_{CR} = 10^{-11}\,\text{s}^{-1}\,n_{H_2}. \qquad (5.35)$$

Balancing the two rates so that $I_{CR} = R_R$ gives

$$x = \frac{3.5 \times 10^{-3}}{\sqrt{(n_{H_2}/m^{-3})}}, \qquad (5.36)$$

and hence the ambipolar diffusion time-scale (see equation 5.31) is given by

$$\tau_{AD} = \frac{2.5 \times 10^{11} \text{ yr}}{\sqrt{(n_{H_2}/m^{-3})}}. \qquad (5.37)$$

This shows that in cloud regions of high density ($\gtrsim 10^{10} \text{ m}^{-3}$), the fractional ionisation will be of order 10^{-8}, and this is so low that the ambipolar diffusion time-scale can be 10^6 years or less. This clearly shows that ambipolar diffusion must be taken into account when studying the effects of magnetic fields in regions of high gas density, and that strict flux freezing cannot be assumed to hold in such regions except on time-scales much shorter than the ambipolar diffusion time.

5.4.3 Decrease of magnetic flux with time

Ambipolar diffusion has very important implications for subcritical clouds supported by magnetic fields. Consider a cloud of initial radius R_0 and density n_{H_2}. As we have seen, the neutrals (which constitute the bulk of the gas mass) will drift inwards at a velocity given by $v = R_0/\tau_{AD}$. This inward drift also increases the central density, thus decreasing the time-scale, and increasing the drift velocity. This has an important effect on the critical mass, as defined in equation 4.82. Note from that equation that

$$M_c \propto B_0 R^2. \qquad (5.38)$$

That is, the critical mass is proportional to the magnetic flux threading the cloud. Now as the neutrals drift past the ions and the field lines, the radius of the cloud decreases but B_0 does not change. So the magnetic flux decreases, and hence the critical mass also decreases. In fact, the magnetic flux (and hence the critical mass) will tend to zero in a time of order (less than) τ_{AD}. So in a time less than τ_{AD}, the actual cloud mass must exceed M_c and the cloud will then start to collapse.

Thus the entire effect of magnetic fields on the equilibrium of clouds turns out to be quite subtle, and can be summarised as follows: Density perturbations which exceed the magnetic critical mass will undergo collapse immediately; for subcritical masses which are sufficiently dense, the magnetic field will hold up collapse for a time given by the ambipolar diffusion time, after which collapse occurs; for subcritical clouds of sufficiently low density and consequently high ionisation, the magnetic field can support the clouds almost indefinitely.

However, in addition to the cloud support role discussed previously, magnetic fields also introduce another important effect into the theory of star formation. They introduce a possible solution to a problem that has been known for a long time, the so-called angular momentum problem.

5.4.4 The angular momentum problem

Consider a cloud of mass M and initial radius R_0, which is rotating with angular velocity ω_0. It has an initial rotational kinetic energy, ignoring factors of order unity, of

$$T_0 \sim M\omega_0^2 R_0^2, \tag{5.39}$$

and an initial gravitational energy

$$\Omega_0 \sim \frac{GM^2}{R_0}. \tag{5.40}$$

The angular momentum, which is conserved, is given by

$$M\omega_0 R_0^2 = M\omega R^2, \tag{5.41}$$

where ω and R are the angular velocity and radius at a later time. Thus as the cloud collapses, R decreases, ω increases, and the rotational kinetic energy increases until it may balance the gravitational energy. In that case a rotationally supported cloud results. The ratio of rotational to gravitational energy at an arbitrary time is given by

$$\frac{T_R}{\Omega_G} = \frac{M\omega^2 R^3}{GM^2} = \frac{M\omega_0^2 R_0^3}{GM^2} \times \frac{R_0}{R} = \frac{T_0}{\Omega_0} \times \frac{R_0}{R}. \tag{5.42}$$

For equilibrium, $T_R/\Omega_G \sim 1$. So the amount of contraction that a cloud will undergo before reaching this equilibrium state depends entirely on the initial ratio of rotational to gravitational energy.

Rotation is an inherent part of motions in the Galaxy. Not only do all clouds orbit about the centre of the Galaxy, but turbulence itself is an inherently rotational phenomenon involving the formation of vortices. Indeed, molecular cloud cores are found to be rotating with typically $T_0/\Omega_0 \sim 0.01$, which is insufficient to stabilise the clouds themselves.

However, this is still a substantial amount of rotational energy. We can see from equation 5.42 that the cloud radius need only decrease by a factor of about 100 before rotation supports the cloud. A typical cloud might start with a radius of 0.1 pc (3×10^{15} m). After it has contracted by a factor of 100 it would still have a radius of 3×10^{13} m. Compare this to the solar radius of $\sim 7 \times 10^8$ m, and it can be seen that this is still five

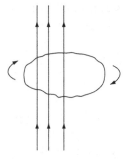 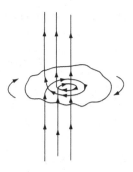

Fig. 5.4. Schematic picture illustrating how cloud core rotation can lead to twisting of the magnetic field lines. Torsional twists travelling along the field lines can carry away excess angular momentum.

orders of magnitude away from forming an object of stellar size! This angular momentum problem is a serious impediment to star formation. Some mechanism needs to be present that allows a rotating cloud to slow down by losing angular momentum, or it could never form a star.

5.4.5 Magnetic braking of rotating clouds

Once it was known that magnetic fields could play important dynamical roles in molecular clouds and cloud cores, it was realised that they could also be an important mechanism for carrying away angular momentum. Intuitively, we expect this since the magnetic fields emerging from a rotating cloud core are tied like rubber bands to the larger Galactic magnetic field, and so would be expected to produce a torque as they are wound up.

Note that this assumes that the rotation axis is parallel to the magnetic field direction. If they are not parallel to one another, then a more complex geometry ensues. However, there are reasons to believe that if they were not initially parallel, then the rotation axis would migrate during the collapse to be parallel to the magnetic field.

As the cloud rotates it starts to twist the field lines – see Figure 5.4. However, as we have seen, such transverse disturbances of the field are known as Alfvén waves (see Section 4.7), which propagate along the field lines at a velocity known as the Alfvén velocity v_A, and would therefore begin to spin-up the lower density gas outside the cloud itself. Since angular momentum must be conserved, this results in a slowing down of the cloud's rotation.

An approximate calculation of the time-scale for this magnetic braking is straightforward. Consider a cloud of mass M, radius R, rotating with angular velocity ω_0, about an axis which is parallel to the prevailing magnetic field B_0. After time t the mass of exterior matter which is co-rotating with the cloud is given by

$$M_{\text{ext}} = \pi R^2 v_A t \rho_{\text{ext}}. \tag{5.43}$$

Since the angular momentum is conserved overall we therefore have that at time t

$$(M + \pi R^2 v_A t \rho_{\text{ext}})\omega(t)R^2 = M\omega_0 R^2, \qquad (5.44)$$

and so finally, the angular velocity is given by

$$\frac{\omega_0}{\omega} = 1 + \frac{\pi R^2 v_A t \rho_{\text{ext}}}{M}. \qquad (5.45)$$

Thus the angular velocity halves in the time it takes for the amount of rotating external gas to equal the mass of the cloud itself. Hence we can calculate a characteristic magnetic braking time t_{MB}, given by

$$t_{\text{MB}} = \frac{M}{\pi R^2 \rho_{\text{ext}} v_A} = \frac{2M}{\pi^{1/2} R^2 \rho_{\text{ext}}^{1/2} B_0}. \qquad (5.46)$$

Thus we see that, by carrying away the angular momentum of a rotating core, the magnetic field can solve the angular momentum problem of star formation.

In this section we have seen the significance of the critical mass, and how even initially subcritical masses can eventually exceed the critical value and undergo contraction. In the next chapter we consider how the cloud evolves after it has exceeded its critical mass. But first we compare the theory of fragmentation we have presented so far in this chapter with observations.

5.5 Observations of the initial conditions of collapse

Observing the initial conditions for collapse and the formation of a star is by definition a tricky undertaking. The formation process takes a long time in human terms, and so we are essentially trying to predict the future evolution of a region in order to say whether or not it will go on to form a star. In so doing we can use the theoretical tools we have developed earlier to test whether a region is gravitationally bound and hence whether it is likely to proceed to form a star.

In searching for these regions we naturally concentrate on the densest cores within molecular clouds. These are variously known as 'starless cores', or (for the most dense cores) 'pre-stellar cores'.[†] Note that for the remainder of this chapter we will concentrate mainly on relatively low-mass stars, i.e. stars of less than a few times the mass of the Sun.

[†] The term 'pre-stellar cores' is an abbreviated form of the original name 'pre-protostellar cores'. They are the same.

(a)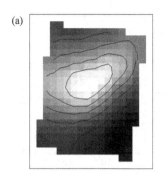

(b)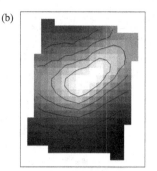

(c)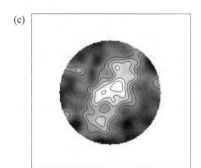

(d)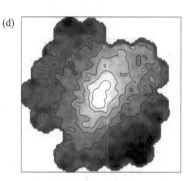

Fig. 5.5. A typical pre-stellar core seen at four different wavelengths from the far-infrared to the submillimetre regime: (a) 170 μm; (b) 200 μm; (c) 450 μm; (d) 850 μm. The resolution and scale of the images are somewhat different. The upper two images cover an area of ~0.6 × 0.4 pc (where 1 pc = 3 × 10^{16} m) with angular resolution equivalent to ~0.05 pc at the distance of the cloud. The lower two images cover a field ~0.1 pc in extent with angular resolution ~0.007 pc, or ~1400 AU (where 1 AU = 1.5 × 10^{11} m), and only show the most dense, inner region of the core, which is a few thousand AU in extent.

5.5.1 Starless and pre-stellar cores

A pre-stellar core is defined as the phase in which a gravitationally bound core has formed in a molecular cloud, and evolves towards higher degrees of central condensation, but no protostar (see Chapter 6) exists yet within the core. The term starless core is a broader category that includes pre-stellar cores, but also includes cores that may not be gravitationally bound. As the name implies, it simply does not contain a star or protostar. Sometimes it is difficult to tell observationally whether or not a starless core will go on to form a star.

Figure 5.5 shows images at different wavelengths of a typical pre-stellar core, which is embedded within the cloud known as Lynds dark cloud 1544 (or L1544 for short). It is shown at wavelengths of 170, 200, 450 and 850 μm, respectively. The images show that the core is not spherically symmetric, but is fairly amorphous. In addition it is not strongly centrally peaked. Even at the highest resolution the densest part of the core still seems to be a few thousand AU in extent.

In fact, pre-stellar cores have density profiles which have a flat inner region, steepening towards the edge. Figure 5.6(a) shows the radial intensity profile of the pre-stellar core L1544 at 850 μm, illustrating this effect. This is similar to the form of radial profile that has been predicted, for example, in magnetically supported cores contracting by ambipolar diffusion.

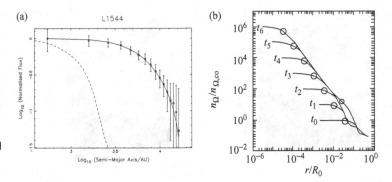

Fig. 5.6. Observed radial profile of pre-stellar core L1544 (left) compared to the theoretical predictions of a model of an initially pressure-supported core, subsequently contracting under self-gravity, moderated by ambipolar diffusion (right). The observed profile is plotted as flux density vs radius on a log-log plot. The dashed line shows how a point source would look on the same plot. The theoretical profiles are plotted as volume density vs radius on a log-log plot at a sequence of predicted times labelled t_0 to t_6, showing how a core is predicted to evolve under this model. The shapes of the theoretical predictions are qualitatively similar to the observations, with a flat inner region of the core and a steeper profile towards the edge.

Figure 5.6(b) shows the theoretical profile of a core, which was originally a pressure-supported sphere, subsequently evolving under ambipolar diffusion. The profiles can be seen to be qualitatively similar to the observed profiles. Hence pre-stellar cores are consistent with some of the theoretical predictions of the star formation models outlined in Section 5.4 above.

5.5.2 Physical properties of pre-stellar cores

Pre-stellar cores emit almost all of their radiation at far-infrared and longer wavelengths, indicating that they must be very cold. Observing this long-wavelength continuum emission allows us to study the dust, and use this as a tracer of the total mass in the cores (see Chapter 2).

The core seen in Figure 5.5 is clearly detected at 170–850 μm. This shows that the core is very cold, and its dust temperature can be obtained by fitting a modified blackbody (sometimes known as a greybody) curve to the observed emission (see Chapter 2). Continuum emission from the core seen in Figure 5.5 can be fitted in this way. Figure 5.7 shows the emission from a typical pre-stellar core. The solid line is a greybody curve of the form

$$F_\nu = B_\nu(T_{dust})\,[1 - \exp(-\tau_\nu)]\,\Omega, \qquad (5.47)$$

where $B_\nu(T_{dust})$ is the Planck function at frequency ν for a dust temperature T_{dust}, τ_ν is the dust optical depth and Ω is the source solid angle (compare this with equation 2.48).

A scaling law for the optical depth is often used, of the form $\tau \propto \nu^\beta$ with $\beta \simeq 1.5$–2. This is found to be the range of values appropriate in this wavelength regime. In Figure 5.7 a good fit is obtained with $T_{dust} = 13$ K and $\beta = 2$. Similar results are obtained in other pre-stellar cores. They have a typical temperature range of $\sim$7–15 K. This confirms

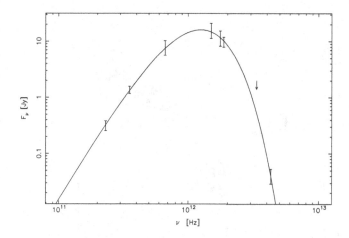

the lack of any warm dust in such cores, and consequently the lack of any embedded star or protostar.

Dust emission is generally optically thin at these long wavelengths, and hence is a direct tracer of the mass content of molecular cloud cores. For an isothermal dust source, the total mass of dust M_d is related to the flux density, F_ν, by the equation (see equation 2.42)

$$M_d = \frac{4a\rho_d F_\nu D^2}{3B_\nu(T_{\text{dust}})Q_\nu}.$$ (5.48)

See Chapter 2 for the derivation of this equation. Following this method, typical masses in low-mass star-forming regions such as Taurus and Ophiuchus ranging from $\sim 0.5 M_\odot$ to $\sim 10 M_\odot$ are derived for different pre-stellar cores.

5.6 Pre-stellar cores and the IMF

In regions of multiple star formation, dust continuum imaging has revealed many cloud fragments, cores and filaments. Figure 5.8 is an 850 μm image of the Ophiuchus molecular cloud showing many cores with characteristic size-scales of ~ 2000–4000 AU,[†] some of which are starless cores and some of which are cores containing embedded stars.

Comparison of the masses derived from the continuum emission with Jeans masses suggests that most of the starless cores are close to gravitational equilibrium and will probably form stars. The typical fragmentation length-scale derived from the average projected separation between cores is ~ 6000 AU in this region. Other regions have different length-scales, depending on their degree of clustering.

[†] 1 AU = 1 astronomical unit = 1.5×10^{11} m.

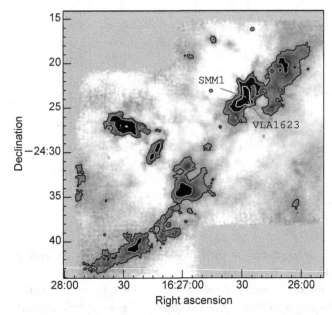

Fig. 5.8. An 850-μm continuum image of the Oph main molecular cloud. The image extent is half a degree, which corresponds to a linear scale of ~1.2 pc at the distance of this cloud. This cloud contains a number of pre-stellar cores and protostars (see Chapter 6). One of each type of core is labelled: the pre-stellar core SMM1; and the protostar VLA1623.

Figure 5.9 shows the mass distribution of the pre-stellar cores seen in Orion. It can be fitted by a function that follows approximately the form (dotted line)

$$\phi(M_{cl})\,dM_{cl} \propto M_{cl}^{-2.3} \; : \; M_{cl} \gtrsim 2.4\mathrm{M}_\odot \tag{5.49}$$

$$\phi(M_{cl})\,dM_{cl} \propto M_{cl}^{-1.3} \; : \; 2.4\mathrm{M}_\odot \gtrsim M_{cl} \gtrsim 1.3\mathrm{M}_\odot \tag{5.50}$$

$$\phi(M_{cl})\,dM_{cl} \propto M_{cl}^{-0.3} \; : \; 1.3\mathrm{M}_\odot \gtrsim M_{cl} \gtrsim 0.4\mathrm{M}_\odot, \tag{5.51}$$

where $\phi(M_{cl})$ for cores is defined in the same way as the IMF for stars and the mass of each core is M_{cl} (see Chapter 1). This pre-stellar core mass function is then seen to resemble the shape of the stellar initial mass function (IMF). Compare these equations with equation 1.5 *et seq.* The power-law indices are the same, although the mass ranges over which they hold true are shifted. So the pre-stellar core mass distribution appears to mimic the shape of the stellar IMF, although there is an offset between the peak mass of the two distributions. This offset probably represents the star-forming efficiency within pre-stellar cores.

This resemblance has also been seen in some other regions. It appears to suggest that the IMF of stars may be determined at the pre-stellar core stage of star formation. We stated in Chapter 1 that the explanation for the stellar IMF appearing universal is one of the chief goals of any star-formation theory. These observations may represent the first clue in trying to understand this phenomenon.

A molecular cloud complex will typically form a cluster of stars. We have been discussing the Ophiuchus star-forming region above. Figure 5.10 now shows a picture of this molecular cloud region, with

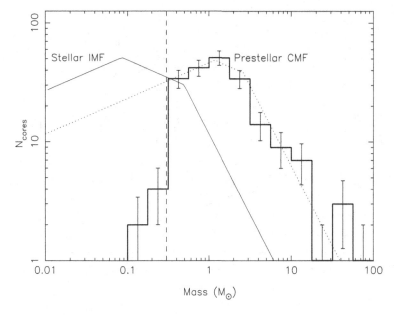

Fig. 5.9. The mass distribution of the pre-stellar cores in Orion. It is plotted simply as a histogram of number versus mass. The histogram can be fitted by three functions very similar to those for the IMF (equation 1.5), which is shown as a thinner solid line. The vertical dashed line shows the point below which the data are incomplete, due to the instrumental sensitivity.

the positions of the newly formed stars marked as crosses. The young stars were detected in the infrared by IRAS (the Infra-Red Astronomical Satellite). The contour marks the approximate edge of the dense molecular cloud material, as traced by the $^{13}CO(1 \to 0)$ emission.

The cluster of recently formed stars is clearly correlated with the position of the molecular cloud. Hence it can be seen that in this case clustered star formation is taking place within this molecular cloud. Thus we can ask how the molecular cloud's core mass spectrum affects the mass distribution of the stars formed. A great deal of interest centres around the IMF of clusters, and indeed of whole galaxies. This is because the IMF of a galaxy determines the evolution of that galaxy. We now briefly discuss binary and multiple star formation.

5.7 Binary and multiple star formation

Roughly two-thirds of all solar-mass main-sequence primary stars are in binary or multiple systems. The primary star is defined as the most massive star in a binary or multiple system.

Furthermore, high-resolution infrared observations of young stars appear to show that most very young solar-mass stars are also in binary systems. Therefore it follows that a significant fraction of stars must be formed in binary systems. There have been many theories put forward to try to explain binary and multiple star formation.

One explanation that was proposed was the capture of one star by the gravitational field of another. This theory has the advantage that it is relatively simple to understand and model. However, it would tend

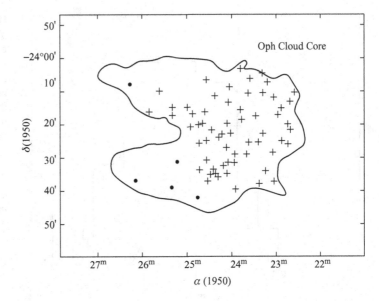

Fig. 5.10. The Ophiuchus molecular cloud region. The crosses mark the positions of the newly formed stars, as detected in the infrared. The contour marks the approximate edge of the dense molecular cloud material.

to produce an increasing binary fraction with age. Such a trend has not been observed. Furthermore, the chance of a close encounter between two stars is too low to produce the observed binary fraction at typical stellar densities in most star-forming regions.

A solution to surmount the latter problem is that discs around the young stars would make interactions between stars more dissipative, thus increasing the frequency of capture. This method may produce some of the observed binary stars, but it has difficulty producing sufficient binaries to match the observations.

One theoretical explanation for the formation of binary and multiple stars is the fragmentation of a collapsing cloud core. Fragmentation occurs as the core collapses, either due to the turbulence within the core, or due to the angular momentum of the core.

If an initially turbulent cloud begins to collapse under its own self-gravity, it will probably fragment in the process, and eventually form a binary or multiple star system. We discussed turbulence and the scale-free nature of interstellar turbulence in Chapter 4. This produces higher-order multiple stars as a natural consequence of the process, as well as binary systems. Furthermore, it also produces binary systems with a large variety of orbital separations and periods. This matches the observations, where there is a wide range of binary separations seen.

If a core, which has a significant initial angular momentum, collapses under its own self-gravity, then it will either form a flattened disc structure, or if it has sufficient angular momentum it will fragment into two or more components. These components can then each go on to

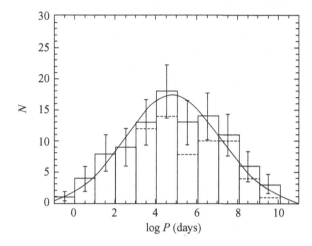

Fig. 5.11. Histogram of number of binary systems versus log(P), where P is the orbital period in days.

form a separate protostar. The result is likely to be a binary or multiple star system.

Another popular explanation for the formation of binary stars is by means of gravitational instability in circumstellar discs. This topic is expanded further in Chapter 8. For now, it can be said that if a gravitational instability occurs in a circumstellar disc, then it can go on to form a star orbiting the primary star. This could be an important mechanism for binary star formation.

Figure 5.11 shows a histogram for binary stars of number of stars observed versus log(P), where P is the orbital period of the binary system in days. The histogram can be approximated by a broad Gaussian distribution – the smooth curve on Figure 5.11. The peak of the Gaussian occurs at around a few times 10^4 days – roughly 200 yrs. This corresponds to a peak orbital separation of roughly 30 AU. The width of the Gaussian extends from less than 1 hr to roughly $\sim 10^5$ yrs.

We note that these statistics were derived for stars similar to the Sun – G-dwarf stars. Other stars show somewhat different statistics. In particular, very low-mass stars show different binary properties. We discuss possible formation mechanisms of very low-mass stars in Chapter 8.

Recommended further reading

We recommend the following texts to the student for further reading on the topics presented in this chapter.

Mannings, V., Boss, A. and Russell, S. (2000). *Protostars and Planets IV*. Tucson: University of Arizona Press.

Reipurth, B., Jewitt, D. and Keil, K. (2007). *Protostars and Planets V*. Tucson: University of Arizona Press.

Ward-Thompson, D. (2002). Isolated star formation: from cloud formation to core collapse. *Science*, **295**, 76–81.

Chapter 6
Young stars, protostars and accretion – building a typical star

6.1 Pre-main-sequence evolution

In this chapter we follow the evolution from a collapsing core in a molecular cloud to a newly formed star as it approaches the main sequence on the Hertzsprung–Russell (HR) diagram. Figure 6.1 sketches the paths followed during the various evolutionary stages on an HR diagram. In this section we briefly outline the various evolutionary stages, and in successive sections we deal with each stage in more detail.

6.1.1 Isothermal collapse

Once a pre-stellar core becomes gravitationally unstable and starts to collapse, then initially the released gravitational energy is freely radiated away and the collapsing fragment stays at roughly the same temperature (isothermal). Its temperature would place it on the right-hand side of the HR diagram (cool), and it has a relatively large radius and hence luminosity. Consequently, it should begin its evolution at the upper right of the HR diagram. Its luminosity is supplied by contraction and the consequent release of gravitational potential energy.

The isothermal collapse phase produces a central concentration of matter and ends with the formation of an opaque, hydrostatic object at the centre, surrounded by a gaseous envelope. We define a hydrostatic object as one which supports itself against gravity by its own internal pressure.

With the increasing density of the gaseous envelope, it becomes increasingly hard for the gravitational potential energy being released

Fig. 6.1. Theoretical
pre-main-sequence tracks on
the Hertzsprung–Russell
diagram.

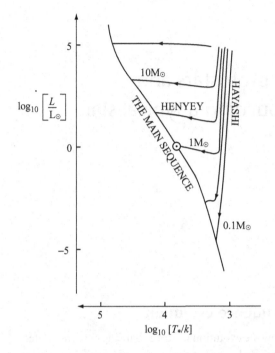

in the interior to diffuse to the outside and escape, and so the luminosity decreases steadily. This part of the evolution is called the Hayashi track. During this phase the interior remains convective. Since the temperature at the outer edge is roughly constant while the luminosity is decreasing, the Hayashi track is roughly vertically downwards on the HR diagram.

The object then enters the main accretion phase during which the central object builds up the majority of its mass ($M_\star$) from a surrounding infalling envelope (of mass M_{env}) and accretion disc, while progressively warming up.

The youngest objects have $M_{env} \gg M_\star$, and radiate by accretion luminosity, given by

$$L_{acc} = \frac{GM_\star \dot{M}}{R_\star}. \tag{6.1}$$

This accretion phase is accompanied by the ejection of a fraction of the accreted material at high velocity in well-collimated flows along two aligned and opposite directions, known as bipolar outflows. These outflows are believed to help carry away the excess angular momentum of the infalling matter, although the outflow mechanism is still not fully understood. When the central object has accumulated most of its final, main-sequence mass, it is known as a pre-main-sequence (PMS) star.

6.1.2 Radiative interior

Eventually the interior becomes hot enough for radiative energy transport to dominate, and it ceases to be convective. It continues to supply most of its luminosity by contracting and releasing gravitational potential energy, but now its luminosity and its surface temperature both rise, and it moves to the left and upwards on the HR diagram, until it reaches the main sequence. This part of the track is called the Henyey track.

When a star reaches the main sequence, the central density and temperature become high enough for hydrogen burning to supply the luminosity, and the star stops contracting. It can be demonstrated that the maximum mass for star formation may be set because, as a star accretes matter, its luminosity increases faster than its mass, and so eventually the force of radiation pressure acting on accreting dust grains reverses the accretion flow.

6.1.3 Protostars and PMS stars

We will continue to use the term pre-stellar core to refer to the initial region of the molecular cloud that starts to collapse, but we here introduce the term 'protostar' to refer to the central hydrostatic object in the middle of the core which forms early in the collapse phase and continues to grow by accretion of the surrounding material to eventually form a star. We refer to the surrounding material as the envelope. Once the envelope has essentially all accreted onto the central protostar (and its circumstellar disc) we will refer to the remaining central object as a pre-main-sequence (PMS) star. We note that these definitions are not unique and the reader may come across different definitions elsewhere.

There are other components of the system that we have not yet discussed. One is the circumstellar disc. This is a by-product of the angular momentum of the accreting material, and is the precursor of planet formation. The other is a bipolar outflow. This is seen in many systems and is high-velocity material flowing away from the protostar along the poles of the system. We discuss circumstellar discs and bipolar outflows in Chapter 8. Figure 6.2 shows a sketch of all of the different components of a protostellar system. It gives the relative positions of the components and their approximate size-scales.

6.2 Hayashi tracks

We now calculate the path that a pre-main-sequence star might follow on the HR diagram.

Fig. 6.2. The different components of a protostellar system.

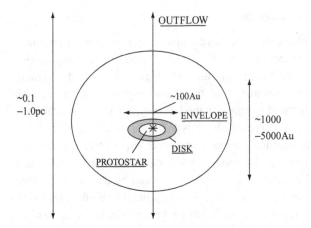

Fig. 6.2. The different components of a protostellar system.

6.2.1 Quasi-static contraction

Once the central protostar becomes too opaque to radiate energy away as fast as energy is being released by gravitational contraction, it rapidly heats up until it is close to hydrostatic equilibrium. At this stage, the main source of opacity is the dust which is mixed in with the gas. After the central protostar has reached approximate hydrostatic equilibrium, it remains there, and thereafter it contracts quasi-statically.

From the virial theorem, we deduce that half of the gravitational potential energy released goes to heat up the gas and maintain approximate hydrostatic equilibrium. We have

$$\frac{k\bar{T}}{\bar{m}} \equiv a^2 \sim v_{escape}^2, \tag{6.2}$$

where $\bar{T}$ is the mean temperature, $\bar{m}$ is the mean particle mass, a is the sound speed within the star and v_{escape} is the escape speed from the surface of the star. Since

$$v_{escape}^2 \equiv \frac{2GM_*}{R} \sim G\,M_*^{2/3}\,\rho^{1/3}, \tag{6.3}$$

the temperature is given by

$$\bar{T} = \frac{G\bar{m}}{k}\,M_*^{2/3}\,\rho^{1/3}. \tag{6.4}$$

The other half of the gravitational potential energy, Ω_G, released supplies the luminosity

$$L_* \simeq \frac{1}{2}\left[-\frac{d\Omega_G}{dt}\right]. \tag{6.5}$$

Quasi-static contraction of a star is sometimes referred to as Kelvin–Helmholtz contraction.

We will now treat the central hydrostatic protostar like any other star and consider its stellar structure. In this context we divide the star into its outer layers, or atmosphere, and its interior. Do not confuse the stellar atmosphere within the outer layers of the protostar with the surrounding envelope of accreting material.

6.2.2 The stellar atmosphere

Eventually the central protostar becomes so hot ($T \overset{>}{\sim} 2000\,\text{K}$) that the dust evaporates. The main source of opacity is then the H^- ion (a proton with two electrons bound to it – albeit that the second electron is only very weakly bound). The mean opacity, $\bar{\kappa}_V$, for the H^- ion is approximately given by

$$\bar{\kappa}_V \simeq \kappa_1 \rho^{3/2} T^{9/2}, \qquad (6.6)$$

where the subscript V indicates that this is the volume opacity coefficient – the total cross-section per unit volume – and κ_1 is a normalisation constant.

We know that in the atmosphere of the protostar, transport of energy has to be by radiation, because ultimately the energy escapes from the star in the form of radiation. Moreover, since most of the gravitational potential energy is released deep down in the interior of the protostar, we know that the flux of radiation through the atmosphere is approximately constant, and we can write

$$F \simeq -\frac{4\,\sigma_{SB}\,T^3}{\bar{\kappa}_V(\rho,\,T)}\,\frac{dT}{dR} \simeq \sigma_{SB}\,T_*^4, \qquad (6.7)$$

where T_* is the star's surface temperature. Equation 6.7 is sometimes termed the radiative equilibrium condition. To solve this equation, we introduce an optical depth variable τ, defined by

$$d\tau = -\bar{\kappa}_V\,dR, \qquad (6.8)$$

$$\Rightarrow \tau(R) = -\int_{R'=R}^{R'=R_*} \bar{\kappa}_V(R')\,dR'. \qquad (6.9)$$

Note that the optical depth τ is measured inwards from the surface of the star. The radiative equilibrium condition now reduces to

$$F \simeq -\sigma_{SB}\,\frac{d}{d\tau}\left[T^4\right] \simeq \sigma_{SB}\,T_*^4, \qquad (6.10)$$

which has the solution

$$T^4(\tau) = T_*^4\,(\tau + 1). \qquad (6.11)$$

This gives us the variation of temperature in the stellar atmosphere, and we see that the temperature increases inwards from $T = T_*$ at the surface, where $\tau = 0$.

Hydrostatic balance in the atmosphere requires that

$$\frac{dP}{dR} = -\frac{G M(R) \rho(R)}{R^2} \simeq -\frac{G M_* \rho(R)}{R_*^2}, \tag{6.12}$$

where we have put $M(R) \to M_*$ and $R \to R_*$, because the atmosphere is only a very thin layer at the surface of the protostar. If we divide equation 6.12 by $\bar{\kappa}_V(\tau) \equiv \bar{\kappa}(\rho(\tau), T(\tau))$, it becomes

$$\frac{dP}{d\tau} = -\frac{1}{\bar{\kappa}_V(\tau)}\frac{dP}{dR} \simeq \frac{G M_* \rho(\tau)}{R_*^2 \bar{\kappa}_V(\tau)}. \tag{6.13}$$

Substituting for the mean opacity from equation 6.6, and for the density ρ from the ideal gas equation of state

$$P = \frac{\rho k T}{\bar{m}} \tag{6.14}$$

$$\implies \rho = \frac{\bar{m} P}{k T}, \tag{6.15}$$

we obtain, after some algebra

$$\frac{dP}{d\tau} \simeq \frac{G M_*}{R_*^2 \kappa_1} \left[\frac{k}{\bar{m} P(\tau)}\right]^{1/2} \frac{1}{T^4(\tau)},$$

$$\implies P^{1/2}\frac{dP}{d\tau} \simeq \frac{G M_*}{R_*^2 \kappa_1} \left[\frac{k}{\bar{m}}\right]^{1/2} \frac{1}{T_*^4(\tau + 1)}, \tag{6.16}$$

which can be integrated to give

$$P^{3/2}(\tau) \simeq \frac{3 G M_*}{2 R_*^2 \kappa_1 T_*^4} \left[\frac{k}{\bar{m}}\right]^{1/2} \ln(\tau + 1),$$

$$\implies P \simeq \left[\frac{3 G M_* \ln(\tau + 1)}{2 R_*^2 \kappa_1 T_*^4}\right]^{2/3} \left[\frac{k}{\bar{m}}\right]^{1/3}. \tag{6.17}$$

This tells us how the pressure varies in the stellar atmosphere as a function of optical depth.

6.2.3 The transition point

Equations 6.11 and 6.17 describe how the temperature and the pressure rise as one penetrates into the outer layers of a star. Convection will develop if the temperature gradient required to drive the radiative flux out of the star becomes too steep. Formally, the condition to prevent convection is

$$\frac{d\ln[P]}{d\ln[T]} < \frac{\gamma}{(\gamma - 1)}. \tag{6.18}$$

But, from equations 6.11 and 6.17,

$$\frac{d\ln[P]}{d\ln[T]} \equiv \frac{T(\tau)}{P(\tau)} \frac{dP/d\tau}{dT/d\tau} \simeq \frac{4}{\ln(\tau+1)},\tag{6.19}$$

and

$$\frac{\gamma}{(\gamma-1)} \simeq \frac{5}{2},\tag{6.20}$$

where γ is the local adiabatic exponent (see Chapter 4), and we have adopted the value $\gamma \simeq 5/3$ appropriate for a monatomic gas. It follows that convection develops once

$$\tau > e^{8/5} - 1 \simeq 4,\tag{6.21}$$

which is just below the surface – four photon mean free paths below the surface in fact. By substituting for τ in equation 6.19, we find that at the transition point, the pressure and temperature reach the values P_{trans} and T_{trans}, given by

$$P_{\text{trans}} \simeq \left[\frac{24\,G\,M_*}{10\,R_*^2\,\kappa_1\,T_*^4}\right]^{2/3} \left[\frac{k}{m}\right]^{1/3},\tag{6.22}$$

$$T_{\text{trans}} \simeq 1.4\,T_*.\tag{6.23}$$

This then defines the transition point from radiative to convective energy transport.

6.2.4 The convective interior

The entire interior of the star (inside $\tau \simeq 4$) is convective, so we have

$$\left\{\frac{d\ln[P]}{d\ln[T]}\right\}_{\text{interior}} \simeq \frac{\gamma}{(\gamma-1)} \simeq \frac{5}{2},$$

hence

$$\left\{\frac{P}{T^{5/2}}\right\}_{\text{interior}} = \text{constant}.\tag{6.24}$$

This ratio holds true at the transition point, so

$$\frac{P_{\text{trans}}}{T_{\text{trans}}^{5/2}} = \text{constant} = \frac{\bar{P}}{\bar{T}^{5/2}},\tag{6.25}$$

where $\bar{P}$ and $\bar{T}$ are mean values of the pressure and temperature in the stellar interior. So this ratio remains roughly constant throughout the stellar interior.

In addition, global hydrostatic equilibrium requires that the internal energy is roughly equal to the gravitational energy, so

$$\bar{P} R_*^3 \sim \frac{G M_*^2}{R_*},$$

and

$$\bar{P} \simeq \frac{G M_*^2}{R_*^4}, \tag{6.26}$$

and hence

$$\frac{k \bar{T}}{\bar{m}} \sim \frac{G M_*}{R_*},$$

thus

$$\bar{T} \simeq \frac{G M_* \bar{m}}{k R_*}. \tag{6.27}$$

This gives us the mean pressure and temperature in the interior of the protostar.

6.2.5 The surface temperature

Substituting into equation 6.25 for P_{trans}, T_{trans}, $\bar{P}$ and $\bar{T}$, from equations 6.22, 6.23, 6.26 and 6.27, we obtain (after some algebra) an expression for the surface temperature, T_*, of the protostar on its Hayashi track

$$T_* \simeq \left\{ G^{13} \, [k/\bar{m}]^{17} \, \kappa_1^{-4} \, M_*^7 \, R_* \right\}^{\frac{1}{31}}. \tag{6.28}$$

It follows that the luminosity, which is given by

$$L_* \simeq 4 \pi \, R_*^2 \, \sigma_{\mathrm{SB}} \, T_*^4, \tag{6.29}$$

is

$$L_* \simeq \left\{ G^{52} \, [k/\bar{m}]^{68} \, \kappa_1^{-16} \, M_*^{28} \, R_*^{66} \right\}^{\frac{1}{31}} \sigma_{\mathrm{SB}}. \tag{6.30}$$

Eliminating R_* between equations 6.28 and 6.30, we obtain an expression for the luminosity in terms of the star's mass and surface temperature

$$L_* \propto M_*^{-14} \, T_*^{66}, \tag{6.31}$$

or, in other words

$$\left. \frac{d\ln[L_*]}{d\ln[T_*]} \right|_{M_*} \simeq 66. \tag{6.32}$$

Rearranging

$$T_* \propto M_*^{0.212} L_*^{0.015}, \qquad (6.33)$$

or, alternatively

$$\left. \frac{d\ln[T_*]}{d\ln[M_*]} \right|_{L_*} \simeq 0.212. \qquad (6.34)$$

These equations define the position and evolution of a protostar on the HR diagram. From equation 6.32, it follows that the tracks for a star of given mass are almost vertical on the HR diagram. From equation 6.34 it follows that tracks for more massive stars are hotter (but only slightly hotter) than those for lower-mass stars (see Figure 6.1).

Since the luminosity is supplied by gravitational contraction, the evolution is in the direction of decreasing R_*, and hence of decreasing L_*. Thus stars evolve downwards on the HR diagram on their Hayashi tracks.

6.3 Henyey tracks

6.3.1 Radiative equilibrium

As the protostar continues to contract and heat up, the dominant source of opacity becomes the opacity due to bound–free and free–free transitions, $\bar{\kappa}$, given by

$$\bar{\kappa} \simeq \kappa_2 \, \rho^2 \, T^{-7/2}, \qquad (6.35)$$

where κ_2 is a normalisation constant, and ρ and T are the density and temperature, respectively. This part of the protostellar evolution is known as the Henyey track.

The temperature dependence of the opacity switches rather abruptly from $\bar{\kappa} \propto T^4$ to $\bar{\kappa} \propto T^{-7/2}$, and as the temperature continues to rise, the opacity falls. At the same time the luminosity is decreasing, and the mean internal temperature is increasing, so the star eventually becomes radiative. In other words, the temperature gradient which is required to drive pure radiative energy transport, $dT/dR \propto \bar{\kappa} L / T^3$, is reduced and consequently the gas becomes stable against convection.

For a star in radiative equilibrium, we have

$$L_* \simeq R_*^2 \, \sigma_{\text{SB}} \, T_*^4, \qquad (6.36)$$

and

$$L_* \simeq \frac{R_* \, \sigma_{\text{SB}} \, \bar{T}^4}{\bar{\kappa}}, \qquad (6.37)$$

therefore

$$T_* \sim \bar{T} \, [\bar{\kappa} R_*]^{-1/4} \equiv \bar{T} \bar{\tau}^{-1/4}, \tag{6.38}$$

where these are standard results from stellar structure.

Substituting for the opacity from equation 6.35, for the mean internal temperature from equation 6.4, and taking the density to be roughly $\rho \sim M_*/R_*^3$, we obtain the optical depth in the form

$$\bar{\tau} \sim \bar{\kappa} R_*$$

$$\simeq \kappa_2 \left[\frac{M_*}{R_*^3} \right]^2 \left[\frac{G M_* \bar{m}}{k R_*} \right]^{-7/2} R_*$$

$$\sim \kappa_2 \left[\frac{G \bar{m}}{k} \right]^{-7/2} [M_* R_*]^{-3/2}. \tag{6.39}$$

This gives us the optical depth in the interior of the protostar.

6.3.2 The surface temperature

Substituting from equations 6.4 and 6.39 in equation 6.38, we obtain an expression for the surface temperature of the star on its Henyey track

$$T_* \simeq \kappa_2^{-1/4} \left[\frac{G \bar{m}}{k} \right]^{15/8} M_*^{11/8} R_*^{-5/8}. \tag{6.40}$$

It follows that the luminosity, which is again given by

$$L_* \simeq 4 \pi R_*^2 \sigma_{\mathrm{SB}} T_*^4 \tag{6.41}$$

is

$$L_* \simeq 4 \pi \sigma_{\mathrm{SB}} \kappa_2^{-1} \left[\frac{G \bar{m}}{k} \right]^{15/2} M_*^{11/2} R_*^{-1/2}. \tag{6.42}$$

Eliminating R_* between equations 6.40 and 6.42, we obtain

$$L_* \propto M_*^{22/5} T_*^{4/5}, \tag{6.43}$$

so

$$\left. \frac{d\ln[L_*]}{d\ln[T_*]} \right|_{M_*} \simeq 4/5, \tag{6.44}$$

and

$$\left. \frac{d\ln[L_*]}{d\ln[M_*]} \right|_{T_*} \simeq 22/5. \tag{6.45}$$

From equation 6.44, it follows that the Henyey tracks for a star of given mass are diagonal (from bottom right to top left) on the HR diagram.

From equation 6.45, it follows that the Henyey tracks for more massive stars are above (well above) those for less massive stars (see Figure 6.1).

Since the luminosity is supplied by gravitational contraction, the evolution is in the direction of decreasing R_*, and hence in the direction of increasing T_* and increasing L_* (see equations 6.40 and 6.42), in other words a protostar evolves upwards and to the left on the HR diagram on its Henyey track.

6.3.3 Very massive stars

For very massive stars, the opacity switches to electron scattering,

$$\bar{\kappa} \simeq \kappa_3 \, \rho,$$ (6.46)

where κ_3 is a normalisation constant, and so the mean optical depth is

$$\bar{\tau} = \bar{\kappa} \, R_* \sim \frac{\kappa_3 \, M_*}{R_*^3}.$$ (6.47)

Substituting from equations 6.4 and 6.47 in equation 6.38, we obtain an expression for the surface temperature of a massive star on its Henyey track

$$T_* \simeq \bar{T} \, \bar{\tau}^{-1/4}$$

$$\simeq \frac{G M_* \, \bar{m}}{k \, R_*} \left[\frac{\kappa_3 \, M_*}{R_*^2} \right]^{-1/4}$$

$$\simeq \left[\frac{G \, \bar{m}}{k} \right] \kappa_3^{-1/4} \, M_*^{3/4} \, R_*^{-1/2}.$$ (6.48)

It follows that the luminosity is again

$$L_* \simeq 4 \, \pi \, R_*^2 \, \sigma_{SB} \, T_*^4$$ (6.49)

and so

$$L_* \simeq \sigma_{SB} \left[\frac{G \, \bar{m}}{k} \right]^4 \kappa_3^{-1} \, M_*^3.$$ (6.50)

From equation 6.50, we see that Henyey tracks for the most massive stars should be approximately horizontal on the HR diagram. In other words, for a star of given mass, the star evolves onto the main sequence at approximately constant luminosity (see Figure 6.1).

Since the luminosity is supplied by gravitational contraction, the evolution is in the direction of decreasing R_*, and hence (from equation 6.48) in the direction of increasing T_*.

Consequently we see that the most massive stars simply evolve horizontally to the left on the HR diagram.

6.3.4 The Kelvin–Helmholtz contraction time-scale

As a newly formed star evolves quasi-statically down its Hayashi track (in approximate convective equilibrium), and across its Henyey track (in approximate radiative equilibrium), its luminosity is supplied by the release of gravitational potential energy (see equation 6.5)

$$L_* \simeq \frac{1}{2} \left[-\frac{d\Omega_G}{dt} \right],$$ (6.51)

and the time-scale for evolution is known as the Kelvin–Helmholtz contraction time-scale, t_{KH}, which is given by

$$t_{KH} \simeq \frac{|\Omega_G|}{L_*} \sim \frac{G\,M_*^2}{R_*\,L_*}.$$ (6.52)

As the star contracts onto the main sequence, R_* decreases, and so this time-scale increases, hence the evolution becomes slower and slower. This is one of the reasons why it is much easier to observe pre-main-sequence stars which are approaching the main sequence than protostars which are still on Hayashi tracks. There are many more stars close to the main sequence.

6.4 Accretion onto protostars

So far in this chapter we have studied what happens in the interior and atmosphere of the protostar itself. We now turn our attention to the interaction between the protostar and its surroundings – namely the envelope and the circumstellar disc (see Figure 6.2). We begin by considering the envelope.

6.4.1 Spherically symmetric accretion

Consider an infinitely extended, stationary background medium with uniform density ρ_0, and uniform isothermal sound speed a_0. Now place a point-mass M_* in the medium. Due to its gravity, the point-mass will grow by accreting the background medium. How rapidly accretion occurs will depend on a number of factors, for instance the star's mass M_*, the envelope density ρ_0, the sound-speed in the envelope a_0, and also on the speed with which the point-mass moves relative to the background medium.

Other factors that affect the accretion include: how the isothermal sound speed in the gas changes as it flows towards the point-mass; and what happens to the gas as it approaches the point-mass. Here we consider the simplest possible case. We assume that the point-mass is at rest relative to the background gas. We assume that the accretion is spherically symmetric. We assume that the gas responds isothermally (i.e. $P = a_0^2 \rho$) with sound speed a_0 constant.

We also assume that the self-gravity of the inflowing gas is negligible, so the gravitational field is dominated by the central point-mass. We seek a steady-state solution in which: (a) the flow variables depend only on position and not on time; and (b) the accretion rate, $\dot{M}_*$, is constant. We also assume that the central point-mass grows sufficiently slowly that its increase in mass can be neglected. This form of accretion, with these assumptions, is known as the Bondi accretion problem.

These assumptions are reasonable provided the central point-mass is much less than the Jeans mass (see equation 4.24) in the background medium, namely that

$$M_* \ll M_{\text{Jeans}} \sim \frac{a_0^3}{G^{3/2} \rho_0^{1/2}}, \tag{6.53}$$

which is usually the case for a star in the general interstellar medium.

Acceptable solutions are constrained by the boundary conditions. In particular, we require that at large distances from the central point-mass, the density tends to its background value, and the inflow velocity tends to zero

$$\text{as } r \to \infty, \qquad \rho \to \rho_0 \qquad \text{and} \quad v \to 0. \tag{6.54}$$

We define the radial velocity v to be positive *inwards*.

6.4.2 Bondi accretion

To derive the equation of motion, we consider the time interval $(t, t + dt)$, but keep in mind that we are looking for a steady-state solution – one in which the flow variables ρ and v are functions of r only, and not t. Material which at time t is at radius r with *inward* velocity $v(r)$ will by time $t + dt$ have reached radius $r + dr$, where

$$r + dr = r - v(r)dt,$$
$$\implies dr = -v(r)dt, \tag{6.55}$$

and so its velocity will have become

$$v(r + dr) = v(r) + \frac{dv}{dr}(r)dr,$$

so

$$v(r + dr) = v(r) - \frac{dv}{dr}(r)v(r)dt. \tag{6.56}$$

The *inward* acceleration is given by

$$a(r) = \frac{v(r + dr) - v(r)}{dt}. \tag{6.57}$$

Combining equations 6.56 and 6.57, we obtain

$$a(r) = -\frac{dv}{dr}(r)v(r).$$ (6.58)

This inward acceleration is produced by a combination of the gravitational acceleration due to the central point-mass, which is given by

$$\frac{GM_*}{r^2},$$

and the acceleration due to the pressure gradient in the inflowing gas, given by

$$\frac{1}{\rho(r)}\frac{dP}{dr}(r).$$

Combining these terms with equation 6.58, we obtain

$$-v(r)\frac{dv}{dr}(r) = \frac{GM_*}{r^2} + \frac{1}{\rho(r)}\frac{dP}{dr}(r).$$ (6.59)

Using the relation between pressure and density

$$P(r) = a_0^2\,\rho(r)$$ (6.60)

we obtain

$$\frac{dP}{dr}(r) = a_0^2\frac{d\rho}{dr}(r).$$ (6.61)

Substituting in equation 6.59 this reduces to

$$-v(r)\frac{dv}{dr}(r) = \frac{GM_*}{r^2} + \frac{a_0^2}{\rho(r)}\frac{d\rho}{dr}(r).$$ (6.62)

This is the general equation of motion for a spherically symmetric, steady-state, non-self-gravitating accretion flow.

6.4.3 Variation of flow speed with radius

We now use the steady-state assumption again. For there to be a steady state, the rate at which material flows inwards across any spherical surface must be constant and equal to the rate at which material accretes onto the central point-mass. Hence we have the condition

$$4\pi r^2\,\rho(r)\,v(r) = \dot{M}_*.$$ (6.63)

Taking logs,

$$\log[4\pi] + 2\log[r] + \log[\rho(r)] + \log[v(r)] = \log[\dot{M}_*].$$ (6.64)

Differentiating equation 6.64 with respect to r, we obtain

$$\frac{2}{r} + \frac{1}{\rho(r)}\frac{d\rho}{dr}(r) + \frac{1}{v(r)}\frac{dv}{dr}(r) = 0, \tag{6.65}$$

for a constant accretion rate. Therefore

$$\frac{1}{\rho(r)}\frac{d\rho}{dr}(r) = -\frac{2}{r} - \frac{1}{v(r)}\frac{dv}{dr}(r). \tag{6.66}$$

Substituting equation 6.66 into equation 6.62, we obtain

$$-v(r)\frac{dv}{dr}(r) = \frac{GM_*}{r^2} - a_0^2 \left[\frac{2}{r} + \frac{1}{v(r)}\frac{dv}{dr}(r)\right],$$

$$\implies -\frac{GM_*}{r^2} + \frac{2a_0^2}{r} = \left[v(r) - \frac{a_0^2}{v(r)}\right]\frac{dv}{dr}(r),$$

$$\implies \frac{dv}{dr} = -\left[\frac{GM_*}{r^2} - \frac{2a_0^2}{r}\right]\left[v(r) - \frac{a_0^2}{v(r)}\right]^{-1}. \tag{6.67}$$

This is the nonlinear differential equation which determines the variation of flow speed with radius.

6.4.4 The sonic point

Like many nonlinear differential equations, equation 6.67 has a singular point, where the denominator is zero. dv/dr becomes infinite at the singular point (which cannot be physical), unless the numerator is also zero there.

The denominator is zero when $v = a_0$, or in other words, the flow velocity is equal to the sound speed. This is known as a sonic point, and we denote this radius as R_{son}. We can put

$$v(R_{son}) = a_0,$$

$$\implies v(R_{son}) - \frac{a_0^2}{v(R_{son})} = 0. \tag{6.68}$$

To avoid a singularity the numerator must also be zero. Accordingly, we must have

$$\frac{GM_*}{R_{son}^2} - \frac{2a_0^2}{R_{son}} = 0$$

$$\implies R_{son} = \frac{GM_*}{2a_0^2}. \tag{6.69}$$

This will then stop equation 6.67 from tending to infinity at the sonic point.

Separating the variables r and v in equation 6.67, we obtain

$$\left[v - \frac{a_0^2}{v} \right] dv = \left[-\frac{GM_*}{r^2} + \frac{2a_0^2}{r} \right] dr, \tag{6.70}$$

which can be integrated to give

$$\frac{v^2}{2} - a_0^2 \ln[v] = \frac{GM_*}{r} + 2a_0^2 \ln[r] + a_0^2 \ln[K]. \tag{6.71}$$

The last term on the right-hand side of equation 6.71, ($a_0^2\ln[K]$), is the constant of integration. We have written it in this form for two reasons. First, it must have the dimensions of speed-squared (like all the other terms in the equation), and a_0 is the only constant speed in the problem. Second, another logarithmic term is required, because the sum of all the logarithmic terms,

$$a_0^2 \{ \ln[v] + 2\ln[r] + \ln[K] \},$$

must give the log of a dimensionless quantity. This sum is ($a_0^2\ln[vr^2 K]$). So K must have dimensions $[L^{-3}T^1]$ for the combined quantity $[vr^2 K]$ to be dimensionless.

Rearranging equation 6.71, we have

$$\frac{v^2}{2} - \frac{GM_*}{r} = a_0^2 \ln[vr^2 K]. \tag{6.72}$$

Substituting from equation 6.63

$$r^2 v = \frac{\dot{M}_*}{4\pi\rho}. \tag{6.73}$$

We obtain

$$\frac{v^2}{2} - \frac{GM_*}{r} = a_0^2 \ln\left[\frac{\dot{M}_* K}{4\pi\rho} \right]. \tag{6.74}$$

We can now invoke the boundary conditions to fix K. Remember these are

$$\text{as } r \to \infty, \qquad \rho \to \rho_0 \qquad \text{and } v \to 0. \tag{6.75}$$

In this limit, the left-hand side of equation 6.74 clearly vanishes, and so the right-hand side must vanish too. This requires

$$\frac{\dot{M}_* K}{4\pi\rho_0} = 1, \tag{6.76}$$

$$\Longrightarrow K = \frac{4\pi\rho_0}{\dot{M}_*}. \tag{6.77}$$

Substituting this back into equation 6.72, we obtain

$$\frac{v^2}{2} - \frac{GM_*}{r} = a_0^2 \ln\left[\frac{4\pi\rho_0 r^2 v}{\dot{M}_*} \right]. \tag{6.78}$$

We can rearrange this equation to make the accretion rate the subject of the equation thus

$$\dot{M}_* = 4\pi \rho_0 r^2 v \exp\left[\frac{GM_*}{a_0^2 r} - \frac{v^2}{2 a_0^2}\right].$$ (6.79)

For a given protostar in a particular location, the values of ρ_0, a_0, M_* and $\dot{M}_*$ are fixed. Then equation 6.79 becomes a relation between v and r.

6.4.5 Physically acceptable solutions

In order to visualise the physical content of equation 6.79, and to determine which values of $\dot{M}_*$ give acceptable solutions, we introduce a dimensionless accretion rate, $\dot{M}_{\rm dim}$, where

$$\dot{M}_{\rm dim} = \frac{a_0^3 \dot{M}_*}{\pi \rho_0 G^2 M_*^2},$$ (6.80)

a dimensionless inward radial velocity, $v_{\rm dim}$, where

$$v_{\rm dim} = \frac{v}{a_0},$$ (6.81)

and a dimensionless radius, $r_{\rm dim}$, where

$$r_{\rm dim} = \frac{2 a_0^2 r}{G M_*} \equiv \frac{r}{R_{\rm son}}.$$ (6.82)

Substituting these into equation 6.79 reduces it to

$$\dot{M}_{\rm dim} = r_{\rm dim}^2 v_{\rm dim} \exp\left[\frac{2}{r_{\rm dim}} - \frac{v_{\rm dim}^2}{2}\right].$$ (6.83)

Figure 6.3 shows the variation of velocity $v_{\rm dim}$ with radius $r_{\rm dim}$, for different values of accretion rate $\dot{M}_{\rm dim}$. These curves can be scaled to physical variables for any given combination of ρ_0, a_0, and M_*, using equations 6.80, 6.81 and 6.82.

From Figure 6.3 it is clear that not all the solutions are physically acceptable. To begin with, for intermediate values of $\dot{M}_*$, the solutions do not extend continuously from $r = 0$ to $r = \infty$. These solutions are marked with dotted lines. Additionally, where these solutions do exist, $v(r)$ is double-valued, which is clearly non-physical – at any given radius there cannot be two radial velocities.

For high values of $\dot{M}_*$, the solutions are continuous in $0 \le r \le \infty$, but as $r \to \infty$, v does not go to zero. These solutions are marked with dashed lines. These solutions must also be rejected, because we require that $v \to 0$, as $r \to \infty$.

This leaves the low values of $\dot{M}_*$, for which again the solutions are continuous in $0 \le r \le \infty$. For these solutions $v \to 0$, as $r \to \infty$, as

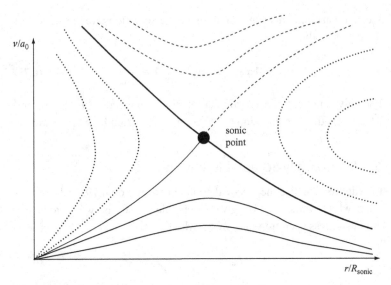

required. Therefore these solutions are physically acceptable. They are
marked with solid lines. The inflow accelerates until it reaches R_{son}, and
then slows down as it approaches the central point-mass.

6.4.6 The supersonic solution

The continuum of acceptable solutions corresponds to a continuum of
accretion rates, $\dot{M}_*$, up to and including a maximum value which corresponds to a solution in which the inflow speed just reaches the speed of
sound before decreasing again as the matter settles towards the centre.
However, because this solution goes through the sonic point, there is
an alternative solution interior to the sonic point, in which the inflow
goes on accelerating to supersonic speeds, and $v \to \infty$ as $r \to 0$. This
solution is marked with a heavy solid line.

This latter solution will probably be the preferred solution in nature,
as it is the one which generates most entropy. Since it goes through the
sonic point, we can substitute $v_{\text{dim}} = 1$ and $r_{\text{dim}} = 1$ in equation 6.83, or
$v = a_0$ and $r = GM_*/2a_0^2$ in equation 6.79, to obtain

$$\dot{M}_{\text{dim}} = \mathrm{e}^{3/2},$$

or, in our normal variables

$$\dot{M}_* = \frac{\mathrm{e}^{3/2}\,\pi\,\rho_0\,G^2\,M_*^2}{a_0^3}. \tag{6.84}$$

This is the critical accretion rate which allows the steady-state flow to
become supersonic ($v > a_0$) near the star, by passing through the sonic
point. It is unique. As the material approaches the centre, it asymptotes

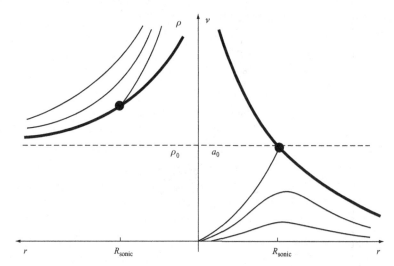

Fig. 6.4. The variation of density (left-hand side) and inward radial velocity (right-hand side) with radius, for the physically acceptable solutions to the Bondi accretion problem.

to free-fall. We note that this result cannot be obtained by dimensional analysis, because there are too many parameters (more than three: G, M_*, ρ_0 and a_0).

Figure 6.4 sketches the variation of density and velocity with radius for the physically acceptable solutions, both those which settle subsonically onto the star, and the one which free-falls supersonically onto the star. Given a solution for $v(r)$, the density can be determined from equation 6.63, i.e.

$$\rho(r) = \frac{\dot{M}_*}{4\pi r^2 v(r)}.$$ (6.85)

6.5 Observations of protostars – the birth line

So far in this chapter we have considered theoretical predictions of how a protostar should evolve towards the main sequence from its beginnings as a pre-stellar core. In the remainder of this chapter we compare some of those predictions with observations.

One of the earliest observational discoveries relating to the Hayashi track was that there were no protostars seen in the top right-hand side of the HR diagram, where theory predicts the start of the Hayashi track to be.

In fact it was noted that there appeared to be a line on the HR diagram, beyond which no protostars were seen. This line is shown on Figure 6.5 and it can be seen to cut directly across the Hayashi track. It is called the birth line.

To understand the origin of the birth line one must remember that the HR diagram is normally plotted in terms of optical colours and luminosities – compare Figure 6.5 with Figure 1.1.

Fig. 6.5. Observations of T Tauri stars on the Hertzsprung–Russell diagram (open circles). The thin line represents the pre-main-sequence track of a 1 $M_\odot$ star. The thick line represents the birth line. Note how the observed stars lie almost exclusively to the left of this line.

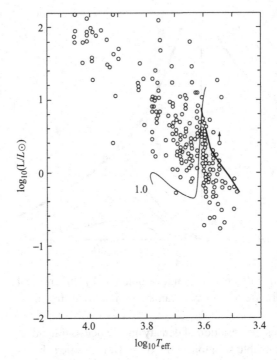

It was soon realised that the reason why no protostars are seen at the top of the Hayashi track is because these are the youngest protostars, and they are therefore surrounded by the most gas and dust. Hence, no optical radiation can escape from the envelopes around these objects as their envelopes are too optically thick.

Consequently, it was realised that to observe the youngest protostars one must observe at longer wavelengths than the optical, where the dust optical depth is lower. In fact, the youngest protostars are only seen at long wavelengths, such as the far-infrared and millimetre-wave regimes.

6.6 Millimetre-wave continuum observations

Surveys carried out to trace millimetre-wave dust continuum emission from molecular clouds have discovered objects which appear to be newly formed, hydrostatic protostars. The youngest of these are designated Class 0 protostars to indicate their youth – see Figure 6.6.

Specifically, Class 0 protostars are defined by the following observational properties: evidence for a central hydrostatic object, as indicated by the detection of a compact centimetre-wave radio continuum source or other indicators of an internal heating source (the radio continuum is interpreted as emission from an accretion shock at the surface of

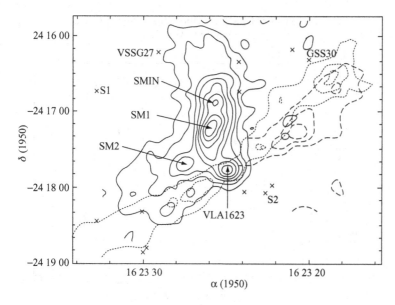

Fig. 6.6. Submillimetre continuum image of the prototype Class 0 protostar designated VLA 1623.

the protostar – cf. Section 5.2); centrally peaked (but slightly extended) millimetre-wave continuum emission, tracing the presence of a circumstellar envelope; a high ratio of millimetre-wave luminosity to total luminosity, suggesting the envelope mass exceeds the central protostar mass (see below); a spectrum resembling a single temperature blackbody at T $\sim$ 15–30 K – see Figure 6.7.

The presence of a compact central object distinguishes Class 0 protostars from the pre-stellar cores discussed in the previous chapter. The other properties distinguish Class 0 protostars from more evolved protostars.

The millimetre-wave luminosity, L_{mm}, is simply proportional to the total dust mass in the protostar's envelope (for a given temperature), and hence to the total envelope mass, M_{env} (see Section 2.8). The total (bolometric) luminosity, L_{bol}, is proportional to the total luminosity of the protostar. This luminosity is provided by accretion onto the central protostar. This can be seen from equation 6.1 to be proportional to the mass of the protostar, $M_\star$, multiplied by the accretion rate.

In the simplest case of a constant accretion rate, L_{bol} is proportional to $M_\star$. Therefore, the ratio of L_{mm}/L_{bol} should roughly track the ratio of $M_{env}/M_\star$. Hence, it is used as an evolutionary indicator (decreasing with time) for protostars. The limiting ratio of L_{mm}/L_{bol} for Class 0 protostars is usually chosen to select objects which have $M_{env}/M_\star \geq 1$.

Class 0 protostars are therefore excellent candidates for being very young accreting protostars in which a hydrostatic core has formed but has not yet accumulated the majority of its final mass. Typical ages of Class 0 protostars are $\sim$ a few $\times 10^4$ up to $\sim 10^5$ years.

Fig. 6.7. Spectral energy distribution of the Class 0 protostar VLA1623.

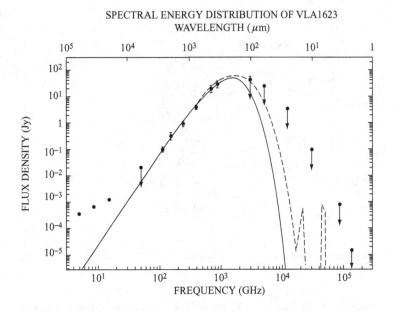

Many Class 0 protostars are multiple systems, sharing a common envelope and sometimes a circumbinary disk. These proto-binary stars probably formed by dynamical fragmentation during the isothermal collapse phase. The discs of Class 0 protostars are typically a factor of ~ 10 times less massive than their surrounding circumstellar envelopes.

6.7 Millimetre-wave spectroscopy

Spectroscopic signatures of infall have been seen towards many Class 0 protostars. Infall motions can be traced by partially optically thick molecular lines, which exhibit asymmetric self-absorbed profiles skewed to the blue. Figure 6.8 shows a typical infall profile. Infall profiles are most often observed in rotational lines of molecules, or molecular ions such as HCO^+.

The reason why infall produces this kind of profile can be understood as an extension of the analysis discussed in Chapter 3. There we showed how a spectral line seen in emission, with a high optical depth, has a profile which is flat at the centre of the line, due to saturation. We also showed how a spectral line seen in absorption appears as a trough, when seen against brighter background emission.

If a line is sufficiently optically thick, then these two effects can combine to produce what is known as a self-absorbed profile, with two peaks either side of an absorption trough. The trough is caused by foreground material absorbing at the same wavelength as background material is emitting. In this case the absorption is produced by static

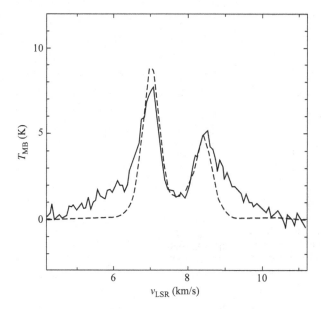

Fig. 6.8. Typical double-peaked spectral line infall profile of a protostar. The blue asymmetry can be clearly seen.

material in the molecular cloud, outside of the collapsing protostellar envelope – see Figure 6.9.

For a collapsing protostellar envelope, the infalling material is moving towards the centre. The nearer half of the envelope is moving away from the observer, and is red-shifted, while the further half of the envelope is moving towards the observer and is blue-shifted. Hence the two peaks can be thought of as tracing the two halves of the infalling envelope – see Figure 6.9.

However, an additional complication arises due to always preferentially seeing material that is closer to the observer. Figure 6.9 shows how on the red-shifted side the observer sees the cooler outer material, further from the protostar, while on the blue-shifted side the observer sees the warmer inner material, closer to the protostar. Warmer material emits more strongly. Hence we see a blue-skewed asymmetric profile associated with infall. The asymmetry can sometimes be produced without the need for a physical temperature gradient if the molecular excitation conditions are right for the particular molecule being studied, but it always indicates some kind of infall.

6.8 Infrared and optical observations

In the infrared, three classes of protostars and pre-main-sequence stars have been identified, based on the spectral slope (or spectral index), $\alpha = d\log(\lambda F_\lambda)/d\log(\lambda)$, of their continuum spectra between 2 and 20 µm.

Fig. 6.9. Simple graphical explanation of how an asymmetric infall profile may arise.

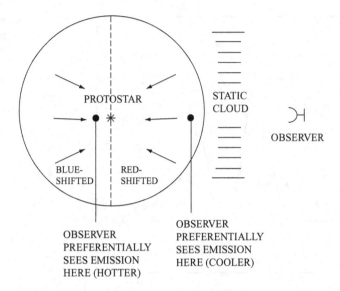

These classes represent an evolutionary sequence that follows on from the Class 0 protostars discussed above.

After Class 0 protostars, the next youngest protostars detected in the infrared are known as Class I sources, and are characterised by $\alpha > 0$. They have typical ages ~ 1–2×10^5 yr and are surrounded by both a disc and a circumstellar envelope of combined mass a few tenths of a solar mass. They derive a substantial fraction of their luminosity from accretion.

The next evolutionary stage is known as the Class II stage. These sources are also known as Classical T Tauri stars (see below) when they are observed in the optical. Class II protostars have $-1.5 < \alpha < 0$. They are surrounded by a circumstellar disc, but have no circumstellar envelope.

The final evolutionary stage of infrared pre-main-sequence stars are known as Class III sources. They have $\alpha < -1.5$, and are also known as weak-line T Tauri stars when observed in the optical (see below). They have no circumstellar envelope, but are surrounded by a remnant circumstellar disc. Figure 6.10 illustrates the different protostellar stages. We now discuss some of the optical properties of Class II and III T Tauri stars.

T Tauri stars are young low-mass stars approaching the main sequence along Henyey tracks. They have masses in the range 0.2–2 $M_\odot$ and luminosities in the range 0.1–20 $L_\odot$. They are found in T associations (see Chapter 1) in molecular clouds. Such stars were first identified because they have strong emission lines in their spectra, in particular Hα. These emission lines are believed to arise in the hot gas accreting onto the central star.

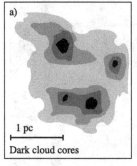

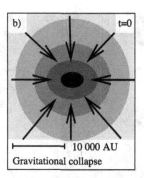

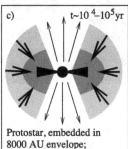

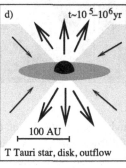

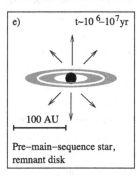

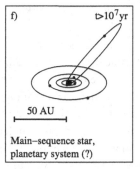

Fig. 6.10. Schematic diagram of the different protostar stages.

T Tauri stars are divided into two types: Classical T Tauri (CTT) stars (alias Class II objects) and weak-line T Tauri (WTT) stars (alias Class III objects). CTT stars have stronger emission lines implying higher levels of accretion from disc to star. WTT stars have weaker emission lines, implying less accretion, and hence presumably a later stage of evolution.

The association of all T Tauri stars with molecular clouds implies that the cloud is the remnant of the material from which the T Tauri stars have recently formed. It is also seen that T Tauri stars have strong infrared emission (see above). This arises from the circumstellar accretion discs. Such discs have radii of a few hundred AU. Estimates of disc masses range from 0.001 to 0.1 $M_\odot$.

These discs have been observed directly by the Hubble Space Telescope. Figure 6.11 shows images of discs around pre-main-sequence

Fig. 6.11. Discs, jets and outflows around young stars.

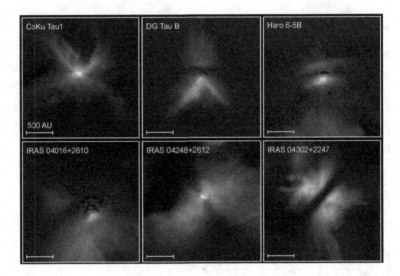

stars. The discs can be seen in absorption in these optical images against the background of diffuse starlight. We return to this topic in Chapter 8.

T Tauri stars have also been detected with X-ray telescopes. The origin of the X-ray emission is not known exactly, but X-ray surveys have more than doubled the number of known T Tauri stars.

T Tauri stars are often seen to be variable in their optical emission. This may be partly due to variable obscuration by a non-uniform density circumstellar disc or envelope. Additionally, time-varying accretion can also cause variable emission. A large amount of material accreting at one time from the inner disc onto the star can lead to an excess of accretion luminosity from the star itself. This can lead to the star brightening for a short time. The most extreme of these events are known as FU Orionis outbursts, after the first star observed to exhibit this phenomenon.

Recommended further reading

We recommend the following texts to the student for further study.

Clemens, D. P. and Barvainis, R. (1994). *Clouds, Cores and Low-Mas Stars*.
 Astronomical Society of the Pacific Conference Series, vol. 65. San Francisco:
 Astronomical Society of the Pacific.

Johnstone, D., *et al.* (2004). *Star Formation in the Interstellar Medium*.
 Astronomical Society of the Pacific Conference Series, vol. 323. San
 Francisco: Astronomical Society of the Pacific.

Stahler, S. W. and Palla, F. (2004). *The Formation of Stars*. Weinheim: Wiley-VCH.

Chapter 7
The formation of high-mass stars, and their surroundings

7.1 Introduction

In this chapter we look at those phenomena associated with the formation of higher-mass stars. High-mass stars are usually defined as stars of mass ~ 8 $M_\odot$ or more. This definition is usually taken, since any star of this mass has typically already begun hydrogen burning before the accretion stage has finished. This provides some problems in dealing with the formation of such stars, since one cannot separate observationally the luminosity due to the accretion, from the intrinsic luminosity of the protostar.

However, the study of high-mass stars is important from the point of view of large-scale studies of galaxies, since the luminosity of a galaxy is typically dominated by the luminosity of its highest mass stars. Hence the observed evolution of a galaxy in terms of its colours and spectra is dominated by the continued formation and evolution of its constituent high-mass stars. Furthermore, high-mass stars are the dominant sources of energy input into the interstellar medium. Hence they are very important for the dynamics and energy budget of a galaxy. In particular, the HII region phase (see below) is particularly important for ionising the gas in the interstellar medium.

Observing high-mass star formation is further complicated by a number of factors. High-mass stars are rarer than low-mass stars, hence the nearest high-mass star-forming regions are on average further away than their low-mass counterparts, making the spatial resolution of observations proportionately lower. The formation and evolution of high-mass stars also occurs much faster than their lower-mass equivalents, making examples of stars in each evolutionary stage rarer. In addition, high-mass stars never appear to form in isolation, but only in clusters, making it more difficult to study the effects and influences of individual stars separately.

Consequently, the theory of the formation of high-mass stars lags behind that for low-mass stars. We still have no detailed theory of the processes involved. Furthermore, the theories that have been proposed are matters of some debate. Hence, we begin by presenting a phenomenological outline of high-mass star formation, based on the empirical phases that are observed. We then go on to present theoretical calculations on the best-studied phase of high-mass star formation, namely HII regions.

7.2 The main stages of high-mass star formation

High-mass stars follow a somewhat different evolution from low-mass stars. There appear to be (at least) three main evolutionary stages that newly forming high-mass stars undergo: infrared-dark clouds; hot molecular cores; and HII regions. This latter stage is itself subdivided into compact HII regions (including hyper- and ultra-compact HII regions), and classical HII regions.

7.2.1 Infrared-dark clouds

The earliest observed phase of high-mass star formation is the infrared-dark cloud (IRDC) stage. IRDCs are very dense, massive, inter-stellar clouds. In fact, they are so dense that they are even optically thick at infrared wavelengths of ~ 1–10 μm. They are typically more dense, and of higher mass, than the clouds from which lower-mass stars form. Little or no detectable star formation has typically begun in these clouds.

The mass of an infrared-dark cloud can be hundreds, or even thousands of solar masses. Often, dense, dark cores can be seen within each cloud. These are known as infrared-dark cores. These cores can have masses of up to ~ 100 $M_\odot$, and are typically only 0.1 pc or less in radius. Hence they have densities of up to $\sim 10^{12}$ hydrogen molecules m^{-3}. These are the most likely known sites of high-mass star formation.

Figure 7.1 shows a typical IRDC. The grey-scale image shows the emission at the infrared wavelength of 8 μm. A dark region can be seen in the lower right of the image. The contours show emission at the much longer wavelength of 850 μm. They can be seen to reach a peak exactly where the 8-μm emission is dark. This means that the dark region is much colder than its surroundings, since colder objects emit at longer wavelengths. Typical temperatures of IRDCs are around 10–20 K. It also shows that there must be a high degree of extinction

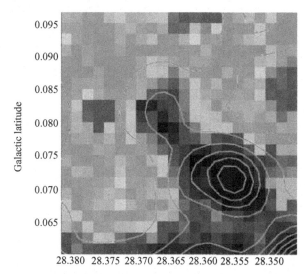

Fig. 7.1. An image of an infrared-dark core. The grey-scale shows emission at 8 μm. The contours show emission at 850 μm. Notice the dark area in the 8-μm emission. This is an infrared-dark core. The contours of 850-μm emission peak at this point, showing that the core emits at the very much longer submillimetre wavelengths.

towards this region, blocking out background starlight at 8 μm. The shorter wavelength emission is absorbed by the dust grains in the cloud and re-emitted at longer wavelengths.

7.2.2 Hot cores

Hot molecular cores are dense interstellar clouds in which the star-formation process has just begun. They are believed to have evolved from infrared-dark clouds after the clouds have begun to be heated by the newly forming stars within them. They exhibit complex molecular chemistry because the heating process initiated by the star formation warms the dust grains within the clouds and evaporates much of the grain mantle material into the gas phase (see Chapter 4). Hot cores have similar masses to IRDCs, but have much higher temperatures and luminosities, which can exceed 10^4 L$_\odot$.

Figure 7.2 shows infrared images of some hot cores. Unlike the IRDCs, these are bright at wavelengths around 8 μm. Hence their temperatures are much higher, as their emission peaks at much shorter wavelengths. Typically they lie in the range of ∼100–200 K. The highest resolution images detect multiple bright objects in the centres of the hot cores. These are believed to be newly forming stars that are heating up the cores.

Hot cores are rich in molecular species that have been released from the surfaces of the dust grains. For example, many carbon chain molecules are seen, including C^+, HCO^+, C_2H_2, C_3H^+, $C_6H_7^+$ and HC_9N, among many others. These molecules formed on the dust grains

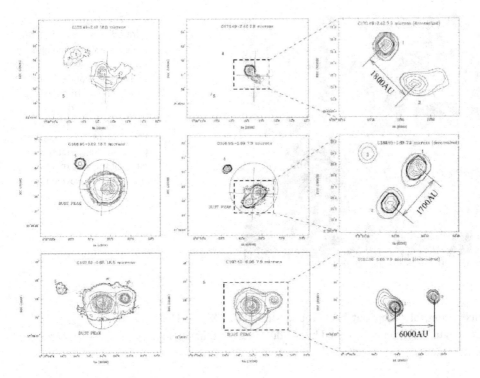

Fig. 7.2. Pictures of three hot cores seen in the mid-infrared at 18.5 μm (left-hand column) and 7.9 μm (middle and right-hand columns). Contours represent brightness. The right-hand column shows enhanced resolution images of the middle column, using a technique which attempts to improve the resolution of the data.

during the IRDC phase and were subsequently released into the gas phase when the newly formed stars heated their surroundings and evaporated the dust grain mantles. When the stars are sufficiently hot and luminous they begin to ionise their surroundings and the next phase begins.

7.2.3 HII regions

HII regions are sharply defined regions of photoionised gas. The ionising photons normally come from O- and B-type stars, which are young, massive stars. Such stars are both extremely luminous ($L_* \stackrel{>}{\sim} 10^4 L_\odot$), and extremely hot, with surface temperatures $T_* \stackrel{>}{\sim} 20\,000$ K.

Therefore, they emit large numbers of energetic photons which ionise and then (following recombination) re-ionise the surrounding hydrogen gas. In fact, most of the ionising photons emitted by the OB stars in an HII region are used maintaining ionisation against recombination, rather than ionising neutral gas for the first time.

The archetypal HII region is the Orion Nebula. There are larger and more luminous HII regions, but being close (at a distance of ∼400 pc) Orion is the best observed. The ionised gas is predominantly hydrogen, so the ionised region is known as an HII region (remember that HI is neutral atomic hydrogen, and HII is singly ionised hydrogen, or H$^+$).

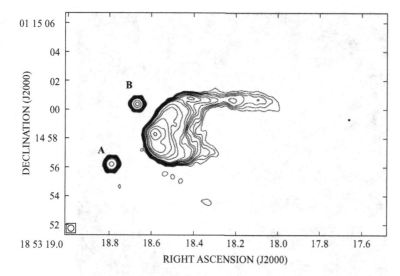

Fig. 7.3. A radio image of a group of HII regions. There is an extended classical HII, as well as two compact HII regions marked 'A' and 'B'.

Compact HII regions are small areas of ionised gas that surround newly formed high-mass stars. Smaller regions are called ultra-compact HII (UCHII) regions, and the smallest of all are known as hyper-compact HII (HCHII) regions. HCHII regions are typically defined as having sizes ≤ 0.01 pc. They have densities $\geq 10^{12}$ m^{-3}.

UCHII regions are usually taken to have a size of between 0.01 and 0.1 pc. They typically have densities $\geq 10^{10}$ m^{-3} and are generally found around B-type or O-type stars. Compact HII regions have sizes between 0.1 and 0.5 pc. Figure 7.3 shows a radio image of some compact HII regions. When compact HII regions burst out of their surrounding molecular cloud they are seen in the optical as classical HII regions. Clearly the more massive the central star, the sooner an HII region will emerge from its cloud, for a given geometry.

Classical HII regions are typically taken to be anything greater than 0.5 pc in size. Once again, they are areas of ionised gas that surround recently formed high-mass stars such as O stars or clusters of stars. They last for the main-sequence life-time of a typical O-type star. Figure 7.4 shows an optical image of an HII region.

Usually an HII region is ionised by the radiation from several stars, either an OB association, or a subgroup of an OB association (for short, an OB subgroup). The OB subgroup which supplies the radiation ionising the Orion Nebula is sometimes called Ori 1c, and is dominated by the four stars in the Trapezium Cluster.

Since OB stars burn out rather quickly – the main-sequence lifetime of an OB star is $\lesssim 3 \times 10^7$ years – HII regions are concentrated near sites of recent and on-going star formation, in giant molecular clouds, in the spiral arms of disc galaxies. Much of the light which delineates the spiral

Fig. 7.4. An optical image of an HII region. Note how the bright stars that have formed in the centre of the cloud have dispersed the surrounding material.

arms in optical images of external galaxies comes from HII regions, and recently formed OB stars. These HII regions can be considerably larger than Orion, and frequently they are arranged along the spiral arm with semi-regular spacing, almost like beads on a string.

When an OB star (or a subgroup of such stars) first forms, it is usually buried deep inside a giant molecular cloud, within the remains of the accretion envelope from which the star has grown – and may still be growing. Consequently the gas surrounding the OB star is very dense, and the HII region it excites is extremely compact. Such compact HII regions are not normally visible in the optical, due to the dust in the surrounding gas, but they can be detected at radio wavelengths, which are not significantly attenuated by dust.

There is a rather effective thermostat, due to line cooling, operating in HII regions, which keeps the temperature of the gas within at most a factor of 2 of 10^4 K. Thus, the effect of an OB star switching on inside a giant molecular cloud is to create a bubble of gas in which the temperature has suddenly increased by a factor $\sim 10^3$ (from ~ 10 K to $\sim 10^4$ K), and additionally the total number density of particles n_{total}, has increased by a factor ~ 4, due to the conversion of each hydrogen molecule into two protons plus two electrons

$$H_2 \longrightarrow 2p^+ + 2e^-. \tag{7.1}$$

As a result, the pressure,

$$P = n_{\text{total}} kT, \tag{7.2}$$

has increased by a net factor $\sim 4 \times 10^3$, and the HII region expands rapidly, sweeping up a shocked layer of dense neutral gas around its rim.

This layer may eventually accumulate a sufficient column density of cool, neutral gas to become gravitationally unstable and fragment. A new generation of stars will probably then condense out of the gravitationally unstable fragments. If some of these new stars are massive stars which in turn excite new HII regions, the process can repeat itself. This is called sequential, or self-propagating, star formation (see Chapter 1).

In addition, an HII region may expand sufficiently to break out of the giant molecular cloud in which it was formed (see Figure 7.5). It will then become optically visible. In general we expect OB stars not to be born precisely at the centres of symmetric giant molecular clouds, and so the resulting HII regions are likely to break out of their parental GMC on one side. This is what appears to be happening in Orion (see Figure 1.10), where there is a dense molecular cloud behind the ionised nebula, and the ionised gas is streaming towards us and away from the molecular cloud at speeds $v \sim 10$ km s^{-1}.

Eventually, the expansion of an HII region may reduce its density to such a low value that it becomes very faint, and hard to detect. Alternatively, an HII region may disappear if the stars supplying it with ionising radiation stop doing so, and the gas then recombines. At this stage, there may be a supernova explosion (or even a sequence of supernova explosions) since this is the manner in which many OB stars end their luminous life-times.

Mature HII regions (i.e. optically visible ones like Orion) are characterised by a spectrum rich in emission lines, of which the most prominent are the recombination lines of hydrogen, which we discuss below, and certain forbidden lines of O$^+$, O^{++} and N$^+$.

The first two of the above listed evolutionary stages (IRDCs and hot cores) are less well understood physically, whereas HII regions have been studied extensively. Consequently, we first consider how to build a high-mass star, and then go on to discuss the details of the HII region phases of high-mass star formation.

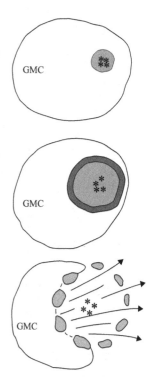

Fig. 7.5. Schematic cross-section of a giant molecular cloud, in which an OB subgroup forms and excites a compact HII region; the HII region then expands under its internal pressure; finally the HII region breaks out of the GMC (cf. Figure 1.11).

7.3 Building a high-mass star

For a high-mass star approaching the main sequence, the Kelvin–Helmholtz time-scale is less than the free-fall time-scale. Consequently, a small hydrostatic hydrogen-burning core forms at the centre, and grows by accretion. In other words, the material that has already reached the central regions of the protostar has time to relax to something very like a main-sequence star, except that it goes on growing in mass and therefore evolves up the main sequence. In the following we make the assumptions that the star accretes spherically symmetrically, and that the accreting medium remains optically thin.

7.3.1 Accretion vs radiation pressure

The inward acceleration due to gravity at distance D from the core, a_{grav}, is given by

$$a_{grav} = \frac{G M_{core}}{D^2}.$$
(7.3)

The outward acceleration due to radiation pressure acting on dust is given by

$$a_{radn} = \frac{L_{core}}{4\pi c D^2} \frac{n_{dust} \sigma_{dust}}{\rho}.$$
(7.4)

Here the first term on the right-hand side $(L_{core}/4\pi c D^2)$, is simply the flux of momentum carried by the radiation from the core, i.e. the radiation pressure. The number of dust grains in unit volume, n_{dust}, is given by

$$n_{dust} = \frac{\rho Z_{dust}}{m_{dust}}.$$
(7.5)

Here Z_{dust} is the dust-fraction by mass, ρ is the total mass in unit volume, which has to be carried along by the radiation pressure force acting on the dust, and m_{dust} is the mass of a single dust grain, given by

$$m_{dust} = \frac{4\pi r_{dust}^3 \rho_{dust}}{3},$$
(7.6)

where r_{dust} is the radius of a dust grain and ρ_{dust} is the density within a dust grain. In addition, σ_{dust}, the effective cross-section presented to radiation pressure by a single dust grain, is given by

$$\sigma_{dust} \simeq \pi r_{dust}^2,$$
(7.7)

providing the radiation has wavelength $\lambda \lesssim 2\pi r_{dust}$.

7.3.2 Reversing the accretion

If the ratio of luminosity to mass, L_{core}/M_{core}, becomes sufficiently large, the outward acceleration due to radiation pressure acting on the dust in the accreting material overpowers the inward acceleration due to gravity.

Combining the above equations, we obtain the condition for accretion to be reversed, in spherically symetric accretion,

$$a_{radn} > a_{grav},$$
(7.8)

and hence

$$\frac{L_{\text{core}}}{4\pi c D^2} \frac{n_{\text{dust}}\,\sigma_{\text{dust}}}{\rho} > \frac{G M_{\text{core}}}{D^2},$$ (7.9)

giving

$$\frac{L_{\text{core}}}{M_{\text{core}}} > \frac{16\pi c G \rho_{\text{dust}} r_{\text{dust}}}{3 Z_{\text{dust}}}.$$ (7.10)

Depending on the properties of the dust, this yields an apparent upper mass limit for main-sequence stars of $\sim$8–20 $M_\odot$. Stars of greater mass than this are seen, so either our assumptions in the above calculation are wrong, or there are other processes at work to build a higher-mass star.

The chief assumption which is most likely to be invalid is that of spherically symmetric accretion. Any angular momentum in a collapsing cloud is likely to lead to some form of flattened structure, possibly a disc of material. Accretion from such a structure will then preferentially occur in the equatorial plane and hence the net outward force felt due to radiation pressure will be reduced.

Another potentially invalid assumption is that the dust grain population in the vicinity of newly forming massive stars is the same as elsewhere in the interstellar medium. It is possible that the abundance of dust is significantly decreased by dust destruction, either by a shock, or by the radiation from the protostar. The dust grain size distribution may also be different, in that all grain mantles will probably be evaporated in the vicinity of the protostar, and hence dust grains in general may be considerably smaller. These two factors would significantly decrease the effective radiation pressure experienced by the accreting material.

In fact, for the formation of the first generation of stars after the Big Bang, known as Population III stars, there would be no heavy elements and hence no dust. Consequently, these stars could have been very much more massive than high-mass stars formed today. In this case the radiation pressure would act most strongly on the electrons in the infalling ionised gas, via Thomson scattering, and the maximum luminosity to mass ratio in the spherically symmetric case would be $\sim$3 $\times$ 10^4 $L_\odot/M_\odot$. This is known as the classical Eddington limit.

Alternatively, our assumption about the effective depth of the gravitational well into which the accreting material is falling may be invalid. High-mass stars mostly appear to form in clusters of stars, such as the Orion Nebula Cluster. In this case the accreting material could experience the potential well of the whole cluster. This would enhance the inward acceleration due to gravity, since M_{core} is replaced by M_{cluster} in

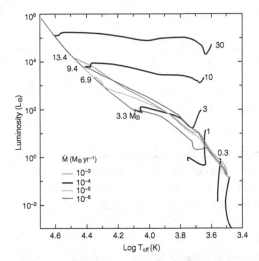

Fig. 7.6. Evolutionary tracks in the Hertzsprung–Russell diagram for non-accreting pre-main sequence stars (thick black lines) and for accreting pre-main sequence stars (grey lines) – see text for details.

equation 7.3. This mechanism could not explain high-mass stars that form in isolation.

Other processes that have been proposed to form high-mass stars include stellar mergers. In this scenario, the sheer density of stars at the centres of clusters is predicted to cause gravitational interactions between neighbouring protostars. These interactions may lead in extreme cases to protostars actually coalescing and merging together to form a higher-mass star.

7.3.3 Pre-main-sequence evolution with accretion

A low-mass star such as the Sun undergoes pre-main-sequence evolution as described in Chapter 6, and arrives at the main sequence when hydrogen burning begins. It then remains on the main sequence at approximately the same point for the majority of its life-time, because its mass doesn't change much.

A high-mass star also reaches the main sequence when hydrogen burning begins. However, the difference for a high-mass star is that it is probably still accreting at the point when it begins to burn hydrogen. Consequently, its mass continues to grow even after it has reached the main sequence, and it evolves along the main sequence.

Figure 7.6 shows some theoretical pre-main-sequence evolutionary tracks for stars of different masses. The black lines show the predicted evolution for protostars that are not accreting, labelled according to the protostar's mass. The grey lines show the predicted evolution for stars which all start at the same initial protostellar mass ($0.1 \, M_\odot$), but which all accrete at differing rates from 10^{-3} to $10^{-6} \, M_\odot \, \text{yr}^{-1}$.

The predicted tracks of the accreting protostars all reach the main sequence at different points, depending on when their hydrogen burning commences. However, thereafter they track along the main sequence as they continue to grow in mass. The tracks shown in Figure 7.6 are somewhat model-dependent, but they give the reader some idea of the different processes that probably have to be taken into account when studying high-mass star formation. The best-studied aspects of high-mass star formation are HII regions. Therefore, we will now look at HII regions in more detail.

7.4 Line radiation from HII regions

The photons of ultraviolet (UV) radiation emitted by a high-mass star can ionise any hydrogen atom with which they interact, provided that their wavelength is less than 91.2 nm (corresponding to a photon energy in excess of 13.6 eV[†]), which is known as the Lyman limit. In this case the electron is removed from the atom, creating an HII ion (i.e. a proton) and a free electron. The HII region extends as far as the luminosity of the star permits, to a radius known as the Strømgren radius. Within this radius protons and free electrons continually interact and recombine to form neutral hydrogen atoms, and photons continually ionise the atoms to create protons and free electrons.

7.4.1 Recombination cascades

The ionised gas in an HII region is continually recombining by means of the following reaction

$$p^+ + e^- \longrightarrow HI + \gamma_\nu, \tag{7.11}$$

where γ_ν represents the emission of a high-energy photon. Not all recombinations go straight into the ground state. About two out of every three recombinations go initially into an excited state. This is usually followed by a cascade of radiative de-excitations down to the ground state, and then re-ionisation from the ground state. Re-ionisation from an excited state is very rare under normal circumstances, because the cascade to the ground state occurs very quickly.

In the interstellar medium, and even in very compact HII regions, the density is much lower than in the best possible laboratory vaccuum. Therefore the separations between nearest-neighbour particles are large. For instance, in an HII region with electron density $n_e \sim 10^9$ m^{-3}, the mean separation between nearest-neighbour particles is $\sim 10^6$ nm.

[†] 1 electron-volt (eV) $= 1.602 \times 10^{-19}$ J.

Consequently, energy levels of atomic hydrogen having large principal quantum number n_{quant}, and hence large extent are well defined and can be occupied. The cross-sectional radius r_n depends on the principal quantum number according to the following relation

$$r_n = \frac{h^2 n_{\text{quant}}^2}{4\pi^2 m_e e^2} \simeq 0.05 \, \text{nm} \, n_{\text{quant}}^2. \tag{7.12}$$

In contrast, at higher densities the electron clouds of these high-n_{quant} levels would overlap. Therefore these high-n_{quant} levels – which are defined by considering a single *isolated* hydrogen atom – do not exist in a high-density gas.

In addition, the selection rules governing spontaneous radiative transitions in excited hydrogen favour transitions in which the principal quantum number changes by a small amount; transitions with $\Delta n_{\text{quant}} = -1$ are preferred over those with $\Delta n_{\text{quant}} = -2$, which in turn are preferred over those with $\Delta n_{\text{quant}} = -3$, etc.

Thus, when an electron/proton pair recombines into an excited state of high principal quantum number, the ensuing cascade to the ground state normally entails the emission of a large number of line photons with frequencies ranging from the radio right through to the ultraviolet. We emphasise that there is a cycle involved here. Electron/proton pairs repeatedly recombine to form hydrogen atoms. Then the hydrogen atom is re-ionised by UV continuum photons from the central star.

At any time, in a typical HII region, the fraction, f_H, of hydrogen in the atomic form is small compared with the fraction which is ionised; and the fraction of atomic hydrogen in excited states is small compared with the fraction which is in the ground state, waiting to be re-ionised

$$f_{\text{excited}} \ll f_{\text{ground-state}} \ll f_{\text{ionized}}. \tag{7.13}$$

We represent the cycle by a sequence of reactions. There is radiative recombination

$$X_f^+ + e^- \longrightarrow X_k + \gamma_\nu \overset{>}{\sim} 10^{10} \, \text{s}, \tag{7.14}$$

which produces continuum radiation (the time-scales on the right-hand side of this and the subsequent equations are estimates for conditions in the Orion Nebula). Remember that X can represent any species, but in the majority of cases is just hydrogen. Then there is the radiative de-excitation cascade

$$\left. \begin{array}{l} X_k \longrightarrow X_j + \gamma_\nu; \\ X_j \longrightarrow X_i + \gamma_\nu; \\ X_i \longrightarrow X_g + \gamma_\nu; \end{array} \right\} \overset{<}{\sim} 0.1 \, \text{s}, \tag{7.15}$$

which produces multiple line emission, since the photon represented by γ_ν has a different frequency in each case. Finally, there is photo-ionisation

$$X_g + \gamma_{\nu_{Lyc}} \longrightarrow X_{f'}^+ + e^- \overset{\gtrsim}{} 10^7 \text{ s}, \qquad (7.16)$$

which produces continuum absorption. In the above, g is the ground state, i, j, k are excited bound levels, and f, f' are free states of the electron/proton pair (see Figure 7.7).

Suppose that we are observing a particular recombination line due to the transition $j \longrightarrow i$. There are hundreds of levels into which the initial recombination can occur, and from each of these there is a multitude of paths to the ground state. Not all of these paths will lead through the transition $j \longrightarrow i$ which we are interested in.

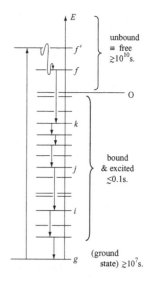

Fig. 7.7. The cycle of recombination, de-excitation and ionisation in an HII region.

7.4.2 Nomenclature

Astronomers use a rather clumsy nomenclature for recombination lines, which reflects the history of their discovery. We should keep in mind that the first recombination lines observed were lines of hydrogen observed in absorption in stellar spectra, and they were one of the principal motivations for the development of quantum mechanics. With the benefit of hindsight, we might devise a better nomenclature.

The various recombination lines of hydrogen are grouped in series. Each series corresponds to a particular lower level, i.e. a particular quantum number n_{lower}. Most of the series are named after one of the physicists involved in their discovery.

The series for which the lower level is the ground state ($n_{lower} = 1$) is called the Lyman series ('Ly' for short). The individual lines are called

Lyα,	$n_{upper} = 2 \to n_{lower} = 1$,	$\lambda = 121.5$ nm;
Lyβ,	$n_{upper} = 3 \to n_{lower} = 1$,	$\lambda = 102.5$ nm;
Lyγ,	$n_{upper} = 4 \to n_{lower} = 1$,	$\lambda = 97.2$ nm;
etc.		

These lines all fall in the UV.

The series for which the lower level is the $n_{lower} = 2$ level is called the Balmer series. The individual lines are called

Hα,	$n_{upper} = 3 \to n_{lower} = 2$,	$\lambda = 656.2$ nm;
Hβ,	$n_{upper} = 4 \to n_{lower} = 2$,	$\lambda = 486.1$ nm;
Hγ,	$n_{upper} = 5 \to n_{lower} = 2$,	$\lambda = 434.0$ nm;
etc.		

These lines fall in the optical, and were therefore the first series to be observed. Being the first, they were simply labelled Hydrogen-α etc., or Hα for short.

The next series of hydrogen recombination lines is called Paschen ($n_{\text{lower}} = 3$), then Brackett ($n_{\text{lower}} = 4$), then Pfund ($n_{\text{lower}} = 5$), and so on. These series fall at successively longer wavelengths.

The convention used in naming other recombination lines is to give (i) the element symbol, (ii) the principal quantum number of the lower level, and (iii) a Greek letter to designate the change in the principal quantum number, according to

$$\Delta n \equiv n_{\text{upper}} - n_{\text{lower}} = \begin{cases} 1 & \alpha \\ 2 & \beta \\ 3 & \gamma \\ 4 & \delta \\ \text{etc.} \end{cases} \tag{7.17}$$

Thus H109α is the line of hydrogen involving a transition from $n_{\text{upper}} = 110$ to $n_{\text{lower}} = 109$; this line falls at $\lambda \simeq 6$ cm. C165β is the line of carbon involving a transition from $n_{\text{upper}} = 167$ to $n_{\text{lower}} = 165$; this line falls at $\lambda \simeq 11$ cm. He$^+$143γ is the line of singly ionised helium which involves a transition from $n_{\text{upper}} = 146$ to $n_{\text{lower}} = 143$; this line falls at $\lambda \simeq 4$ cm. These are all radio recombination lines.

Radio recombination lines are particularly important because they do not suffer from significant dust obscuration, and so they can be used to measure radial velocities (using the Doppler effect) on the far side of the Galaxy. This in turn allows us to map out the distribution and kinematics of ionised gas throughout the disc of the Galaxy.

7.5 Recombination rate and emission measure

The recombination rate, $\mathcal{R}_X$, per unit volume, for species X is given by

$$\mathcal{R}_X = \alpha_X(T) n_{X^+} n_e, \tag{7.18}$$

where $\alpha_X(T)$ is the recombination coefficient at temperature T, and n_{X^+} and n_e are the numbers per unit volume of ionised species X^+ and electrons respectively. From now on we will concentrate mainly on hydrogen, which has

$$\alpha_{\text{H}}(T) \simeq 3 \times 10^{-16}\, \text{m}^3\, \text{s}^{-1}\, (T/\text{K})^{-3/4}. \tag{7.19}$$

If the fraction of all recombinations which leads through the transition $j \rightarrow i$ is $f_{ji}(T)$, then the integrated volume emission coefficient for the line $j \rightarrow i$ is

$$j = \alpha_{\text{H}}(T) n_{\text{p}} n_e f_{ji}(T) \frac{h\nu_0}{4\pi}, \tag{7.20}$$

where we have replaced n_{X^+} with n_p, the number of free protons, for the case of hydrogen. It is found that $f_{\text{H}\alpha} \sim 0.3$, i.e. only about 30% of hydrogen recombinations produce an Hα photon. Similarly $f_{H\beta} \sim 0.1$, so only about 10% of hydrogen recombinations produce an Hβ photon.

If we look at an HII region on a line of sight where there is negligible background intensity, and if the emission from the HII region is optically thin, then (from Chapter 2), we have

$$I_\nu^{obs} \simeq \int_{los} S_\nu(\tau_\nu)\, d\tau_\nu$$

$$\simeq \int_{los} j_\nu(l)\, dl \simeq \int_{los} j(l)\, \phi_\nu(l)\, dl, \tag{7.21}$$

where I_ν^{obs} is the observed intensity, $\int_{los}$ represents an integration along the line of sight with respect to l, and $\phi_\nu(l)$ is the line profile function. It follows that the integrated intensity is

$$I = \int_{line} I_\nu\, d\nu = \int_{los} j(l)\, dl$$

$$= \frac{h\nu_0}{4\pi} \int_{los} \alpha_H(T(l)) f_{ji}(T(l))\, n_p(l)\, n_e(l)\, dl. \tag{7.22}$$

If the emitting medium has uniform temperature $T = T_0$, then

$$I = \frac{h\nu_0}{4\pi} \alpha_H(T_0) f_{ji}(T_0) \int_{los} n_p(l)\, n_e(l)\, dl. \tag{7.23}$$

The integral in the last equation is called the emission measure of the line of sight, $\mathcal{E}M_H$, where

$$\mathcal{E}M_H = \int_{los} n_p(l)\, n_e(l)\, dl. \tag{7.24}$$

If the emitting medium has uniform density, then

$$\mathcal{E}M_H = n_p\, n_e\, L, \tag{7.25}$$

where L is the length of the intercept which the line of sight makes with the emitting medium.

Thus the integrated intensity of an optically thin recombination line enables us to estimate the emission measure of the emitting medium from equation 7.23. Suppose we also have an independent estimate of the linear size L of the emitting region, say from its distance D and its angular size θ, via $L = D\theta$ (assuming spherical symmetry, which is often a good assumption). Then we can combine this with the emission measure to obtain

$$\langle n_p\, n_e \rangle = \frac{\mathcal{E}M_H}{L}. \tag{7.26}$$

Typically in an HII region, we have

$$n_p \sim n_e, \tag{7.27}$$

and so the hydrogen emission measure obtained from a hydrogen recombination line gives

$$\langle n_e^2 \rangle \sim \langle n_p n_e \rangle = \frac{\mathcal{E} M_H}{L}. \tag{7.28}$$

We see that, from the study of hydrogen recombination lines, we can estimate the density of gas in an HII region.

7.6 Free–free radio continuum emission

An accelerated charge emits radiation. When the accelerated charge is a thermal electron (i.e. one of the ordinary electrons with a Maxwellian velocity distribution), and it is accelerated by the Coulomb attraction of a nearby thermal proton, the radiation is called free–free emission, or thermal bremsstrahlung radiation.

We must keep in mind that this emission is produced by an electron/proton pair, not by an isolated electron. The designation 'free–free' simply refers to the fact that the electron/proton pair is in a free state, both before, and after, the interaction which produces the emission. Bremsstrahlung is the German for 'braking radiation', since the emitting electron/proton pair are being braked rather than accelerated.

When the accelerated charge is a relativistic electron (i.e. a cosmic-ray electron), and it is accelerated by the Lorentz force, F_L, given by

$$F_L = -\frac{e\mathbf{v} \times \mathbf{B}}{c}, \tag{7.29}$$

due to the interstellar magnetic field $\mathbf{B}$, the radiation is known as synchrotron emission.

The monochromatic volume emission coefficient for free–free radiation, j_ν, is given by

$$j_\nu = \beta_\nu(T) n_p n_e, \tag{7.30}$$

where, at radio wavelengths

$$\beta_\nu(T) \simeq 3 \times 10^{-51} \, \mathrm{J\,m^3\,s^{-1}\,ster^{-1}\,Hz^{-1}} \left(\frac{T}{K}\right)^{-1/2} \left(\frac{\nu}{Hz}\right)^{-0.1}. \tag{7.31}$$

Note that the volume emission coefficient in equation 7.30 depends on the product of the number densities of the species involved in the two-body interactions which give rise to the free–free photons.

Since the initial and final free states of the electron/proton pair are populated as in thermodynamic equilibrium at temperature T (i.e. there is a Maxwellian distribution of electron and proton velocities at a common temperature T), the source function approximates also to its thermodynamic equilibrium value, i.e. the Planck function (or blackbody function) at temperature T

$$S_\nu \simeq B_\nu(T) \simeq \frac{2kT\nu^2}{c^2}. \tag{7.32}$$

The last expression is the Rayleigh–Jeans approximation to the Planck function, which is only valid when $h\nu \ll kT$.

Combining equations 7.30, 7.31 and 7.32, we obtain an expression for the monochromatic volume opacity coefficient

$$\kappa_\nu = \frac{j_\nu}{S_\nu} \simeq 10^{-11}\,\mathrm{m}^5 \left(\frac{\nu}{\mathrm{Hz}}\right)^{-2.1} \left(\frac{T}{\mathrm{K}}\right)^{-1.5} n_p n_e. \tag{7.33}$$

Hence the optical depth is given by

$$\tau_\nu = \int_{los} \kappa_\nu(l)\,dl$$

$$\simeq 10^{-11}\,\mathrm{m}^5 \left(\frac{\nu}{\mathrm{Hz}}\right)^{-2.1} \int_{los} \left(\frac{T(l)}{\mathrm{K}}\right)^{-1.5} n_p(l)\,n_e(l)\,dl. \tag{7.34}$$

If the line of sight is dominated by an HII region having uniform temperature (T_0), then the last equation reduces to

$$\tau_\nu = 10^{-11}\,\mathrm{m}^5 \left(\frac{\nu}{\mathrm{Hz}}\right)^{-2.1} \left(\frac{T_0}{\mathrm{K}}\right)^{-1.5} \int_{HII} n_p(l)\,n_e(l)\,dl, \tag{7.35}$$

where $\int_{HII}$ implies integration through the HII region. Moreover, the integral in the above equation is the emission measure $\mathcal{E}M_H$ – see equation 7.24. Hence

$$\tau_\nu = 10^{-11} \left(\frac{\nu}{\mathrm{Hz}}\right)^{-2.1} \left(\frac{T_0}{\mathrm{K}}\right)^{-1.5} \left(\frac{\mathcal{E}M_H}{\mathrm{m}^{-5}}\right). \tag{7.36}$$

Thus the emission measure also gives us a handle on the optical depth.

If we look at an HII region with uniform temperature T_0, and the background intensity is negligible, we should observe free–free radiation with intensity given by

$$I_\nu(\tau_\nu) = S_\nu \left[1 - e^{-\tau_\nu}\right] = B_\nu(T_0) \left[1 - e^{-\tau_\nu}\right]. \tag{7.37}$$

We can identify a critical frequency $\nu_{turnover}$, such that the spectrum of free–free radiation has different asymptotic forms for $\nu \ll \nu_{turnover}$ and $\nu \gg \nu_{turnover}$. Here $\nu_{turnover}$ is the frequency at which the optical depth equals unity.

At high frequencies, the emission is optically thin, so we can write, for $\nu \gg \nu_{turnover}$ and $\tau_\nu \ll 1$

$$I_\nu \simeq B_\nu(T_0)\tau_\nu \propto \nu^{-0.1} T_0^{-0.5} \mathcal{E}M_H. \tag{7.38}$$

Fig. 7.8. The free–free continuum spectrum from an HII region, showing the transition from optically thick at $v \ll v_{turnover}$ to thin at $v \gg v_{turnover}$.

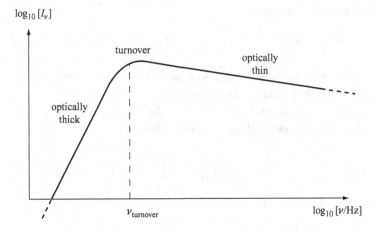

Conversely, at low frequencies the emission is optically thick, so we can write, for $v \ll v_{turnover}$ and $\tau_v \gg 1$

$$I_v \simeq B_v(T_0) \simeq \frac{2kT_0 v^2}{c^2}. \tag{7.39}$$

Figure 7.8 illustrates a typical free–free continuum spectrum from an HII region. We can clearly identify the two asymptotic forms, and the turnover frequency $v_{turnover}$.

From the intensity at low frequencies (where the emission is optically thick), we can obtain an estimate of the gas kinetic temperature

$$T_0 \simeq \frac{c^2 I_v}{2kv^2}, \tag{7.40}$$

for $v \ll v_{turnover}$. Typical HII regions have temperatures $T_0 \sim 10^4$ K and almost always have $7000\,\text{K} \lesssim T_0 \lesssim 14\,000\,\text{K}$.

We can also obtain an estimate of the emission measure from the turnover frequency. The optical depth at the turnover frequency is 1. Hence

$$10^{-11} \left(\frac{T_0}{K}\right)^{-1.5} \left(\frac{v_{turnover}}{Hz}\right)^{-2.1} \left(\frac{\mathcal{E}M_H}{m^{-5}}\right) = 1, \tag{7.41}$$

and therefore

$$\mathcal{E}M_H \simeq 10^{11}\,\text{m}^{-5} \left(\frac{T_0}{K}\right)^{1.5} \left(\frac{v_{turnover}}{Hz}\right)^{2.1}. \tag{7.42}$$

From equation 7.28 we have

$$n_p \sim n_e \sim \left(\frac{\mathcal{E}M_H}{L}\right)^{1/2}, \tag{7.43}$$

and once again we have an estimate of the gas density. Remember that these estimates of n_p and n_e are strictly speaking root-mean-square values averaged along the line of sight.

7.7 Size of an HII region – Strømgren radius

The size of an HII region is also known as the Strømgren radius, and is determined by equating the rate at which the central star (or stars) emits hydrogen-ionising photons, $\dot{\mathcal{N}}_{\mathrm{H}}$, to the rate at which protons and electrons recombine throughout the total volume of the HII region

$$\dot{\mathcal{N}}_{\mathrm{H}} = \frac{4\pi R_{\mathrm{HII}}^3}{3} \alpha_{\mathrm{H}}^*(T) n_p n_e \simeq \frac{4\pi R_{\mathrm{HII}}^3}{3} \alpha_{\mathrm{H}}^*(T) n_e^2. \tag{7.44}$$

Here, R_{HII} is the radius of the HII region, and $\alpha_{\mathrm{H}}^*(T)$ is the recombination coefficient for atomic hydrogen, taking into account only recombinations into excited states, given by

$$\alpha_{\mathrm{H}}^*(T) \sim 2 \times 10^{-16}\,\mathrm{m}^3\,\mathrm{s}^{-1} \left[\frac{T}{\mathrm{K}}\right]^{-3/4}. \tag{7.45}$$

We have obtained the last expression in equation 7.44 by approximating $n_p \simeq n_e$. This is justifiable on the grounds that all elements other than hydrogen are so much less abundant that they cannot contribute many additional free electrons. The next largest contribution is from helium, which could increase n_e by up to 20%, but seldom increases n_e by more than 10%, because in general there are not enough photons with sufficient energy to ionise helium.

It follows from equation 7.44 that

$$R_{\mathrm{HII}} \simeq \left[\frac{3\dot{\mathcal{N}}_{\mathrm{H}}}{4\pi \alpha_{\mathrm{H}}^*(T) n_e^2}\right]^{1/3}, \tag{7.46}$$

which simply embodies the fact that if the density in an HII region is reduced, the rate of recombination per unit volume, $\sim \alpha_{\mathrm{H}}^*(T) n_e^2$, decreases too, and so the ionising radiation from the central star can maintain ionisation against recombination in a larger volume.

In obtaining equation 7.46 we have assumed that all of the ionising radiation from the central star is used up maintaining ionisation against recombination, rather than ionising new material at the edge of the HII region. We shall show later that this is a reasonable approximation. Nonetheless, because the HII region expands due to its over-pressure, the density in the already-ionised gas falls, and hence the recombination rate in this gas falls too. Hence there is always a little ionising radiation left over to ionise new material at the edge of the HII region, and so the total mass of ionised gas,

$$M_{\text{HII}} = \frac{4\pi R_{\text{HII}}^3 n_p m_p}{3 X_{\text{H}}} \simeq \frac{\dot{\mathcal{N}}_{\text{H}} m_p}{\alpha_{\text{H}}^*(T) n_e X_{\text{H}}}, \qquad (7.47)$$

increases with time (X_{H} is the hydrogen mass fraction). The difference between $\alpha_{\text{H}}^*(T)$ and $\alpha_{\text{H}}(T)$, the recombination coefficient for hydrogen (see equation 7.19) is that $\alpha_{\text{H}}^*(T)$ takes into account only recombinations going initially into excited states. In other words it neglects recombinations straight into the ground state. The reason that this approximation can be made is the following.

When a proton and an electron recombine straight into the ground state of atomic hydrogen, an ionising photon is emitted. Typically this ionising photon has energy just above the threshold for ionisation. The cross-section presented by a hydrogen atom to the photon is therefore very large. Consequently, the ionising photon is very likely to be absorbed close to its point of emission, producing a compensatory ionisation. The upshot is that recombinations straight into the ground state do not have to be reversed by photons from the central star, and therefore they need not be taken into the reckoning when we are calculating the radius of the HII region. This is called the 'on-the-spot approximation'.

Substituting $T_0 \sim 10^4$ K, whence $\alpha_{\text{H}}^*(T_0) \sim 2 \times 10^{-19}$ m^3 s^{-1}, and adopting typical reference values, equations 7.46 and 7.47 give the Strømgren radius,

$$R_{\text{HII}} \simeq 1.7 \, \text{pc} \left[\frac{\dot{\mathcal{N}}_{\text{H}}}{10^{50} \, \text{s}^{-1}} \right]^{1/3} \left[\frac{n_0}{10^9 \, \text{m}^{-3}} \right]^{-2/3}, \qquad (7.48)$$

and the mass of the HII region,

$$M_{\text{HII}} \simeq 600 \, \text{M}_\odot \left[\frac{\dot{\mathcal{N}}_{\text{H}}}{10^{50} \, \text{s}^{-1}} \right] \left[\frac{n_0}{10^9 \, \text{m}^{-3}} \right]^{-1}. \qquad (7.49)$$

The reason that the recombination coefficient decreases with increasing temperature (see equation 7.45) is that at higher temperatures the protons and electrons are moving faster. Therefore, although they pass one another more frequently, the time available for them to interact is shorter, and the interaction cross-section is smaller.

7.8 Ionisation fronts

Here we estimate the thickness of the transition region at the edge of an HII region, where the degree of ionisation falls from (say) ~90% to (say) ~10%. We conclude that the transition is usually extremely abrupt, and therefore can reasonably be described as a 'front'.

Suppose that $F_n(r)$ is the radial number flux of hydrogen-ionising photons from the central star. The subscript 'n' is to denote that F_n measures the number of hydrogen-ionizing photons crossing unit area in unit time (rather than the amount of radiant energy crossing unit area in unit time). Suppose also that the HII region has approximately uniform temperature,

$$T(r) \simeq T_0 \sim 10^4 \, \text{K}, \tag{7.50}$$

and approximately uniform density, n_0, so that

$$n_e \simeq n_p \simeq n_0 - n_{\text{HI}}. \tag{7.51}$$

In putting $n_e \simeq n_p$, we have again neglected contributions to the electron density n_e from the ionisation of elements other than hydrogen.

The equation of ionisation balance equates the rate of hydrogen ionisation per unit volume, $\mathcal{I}_{\text{HI}}$, to the rate of hydrogen recombination per unit volume, $\mathcal{R}_{\text{HI}}$, namely

$$\mathcal{I}_{\text{HI}} \simeq F_n(r) \, n_{\text{HI}}(r) \, \bar{\sigma}_{\text{HI}} \simeq \mathcal{R}_{\text{HI}} \simeq \alpha^*_{\text{HI}}(T_0) \, n_e(r) \, n_p(r), \tag{7.52}$$

where $\bar{\sigma}_{\text{HI}}$ is the mean cross-section presented by a hydrogen atom to an average ionising photon from the central star, $\bar{\sigma}_{\text{HI}} \sim 7 \times 10^{-22} \, \text{m}^2$.

Next we substitute

$$n_e(r) = n_p(r) = x(r) n_0, \tag{7.53}$$

and

$$n_{\text{HI}}(r) = (1 - x(r)) \, n_0, \tag{7.54}$$

where $x(r) = n_e(r)/n_0$ is the degree of ionisation at radius r. Equation 7.52 then reduces to

$$F_n(r) = \frac{\alpha^*_{\text{HI}}(T_0) n_0}{\bar{\sigma}_{\text{HI}}} \frac{x^2(r)}{[1 - x(r)]}. \tag{7.55}$$

We shall also need the result

$$\frac{dF_n}{dr} = \frac{\alpha^*_{\text{HI}}(T_0) n_0}{\bar{\sigma}_{\text{HI}}} \frac{x(r)[2 - x(r)]}{[1 - x(r)]^2} \frac{dx}{dr} \tag{7.56}$$

for later.

We concentrate first on the interior of the HII region, and consider the spherical shell-element between r and $r + dr$. The rate at which hydrogen-ionising photons flow across the spherical surface with radius r is $4\pi r^2 F_n(r)$, and so the rate at which hydrogen-ionising photons are used up within the element is given by

$$d\mathcal{N}_n^1 = -\frac{d}{dr} \left[4\pi r^2 F_n(r) \right] dr. \tag{7.57}$$

The rate at which hydrogen-ionising photons are used up within the element maintaining ionisation against recombination is also given by the volume of the element times the rate of recombination per unit volume

$$d\mathcal{N}_n^2 = 4\pi r^2 dr \left[\alpha_{\rm HI}^*(T_0)\, n_e(r)\, n_p(r)\right].$$

(7.58)

Equating 7.57 and 7.58, which both give the rate at which hydrogen-ionising photons are used up in the element, and cancelling the common factor dr, we obtain the equation of radiative transport for the flux of hydrogen-ionising photons, in the form

$$\frac{d}{dr}\left[4\pi r^2 F_n(r)\right] = -4\pi r^2 \left[\alpha_{\rm HI}^*(T_0)\, n_e(r)\, n_p(r)\right].$$

(7.59)

If we anticipate the result that in the interior of the HII region the degree of ionisation is very high, i.e. $x \sim 1$, and approximate

$$n_e(r) = n_p(r) = n_0,$$

(7.60)

then the equation of radiative transport reduces to the form

$$\frac{d}{dr}\left[4\pi r^2 F_n(r)\right] = -4\pi r^2 \left[\alpha_{\rm HI}^*(T_0)\, n_0^2\right],$$

(7.61)

which is readily integrated to give

$$\left[4\pi r^2 F_n(r)\right] = -\frac{4\pi r^3 \alpha_{\rm HI}^*(T_0)\, n_0^2}{3} + \dot{\mathcal{N}}_{\rm HI}.$$

(7.62)

The last term is the constant of integration, which ensures that as $r \to 0$ the flow of hydrogen-ionising photons across r approaches the output of ionising photons from the central star

$$\text{Limit}_{r\to 0}\left\{4\pi r^2 F_n(r)\right\} = \dot{\mathcal{N}}_{\rm HI}.$$

(7.63)

Rearranging equation 7.62, we obtain

$$F_n = \frac{\dot{\mathcal{N}}_{\rm HI}}{4\pi r^2} - \frac{\alpha_{\rm HI}^*(T_0)n_0^2 r}{3},$$

(7.64)

where the first term on the right-hand side represents the unattenuated flux from the central star falling off as r^{-2}, and the second term represents the attenuation due to absorption of the hydrogen-ionising photons.

Turning now to the edge of the HII region, we can no longer assume that $x \sim 1$, because in this region x falls to values $x \ll 1$. However, anticipating our final result, we can assume that the transition region (or ionisation front, IF) is very thin, as compared with the overall radius of the HII region

$$\Delta R_{\rm IF} \ll R_{\rm HII},$$

(7.65)

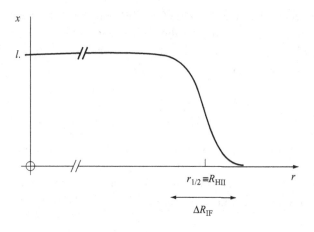

Fig. 7.9. The variation of the degree of ionisation x with radius r, in the vicinity of the boundary of an HII region.

and so

$$r \simeq R_{\text{HII}}. \tag{7.66}$$

With this assumption, we can neglect the curvature of the transition region, and reduce the equation of radiative transport for the hydrogen-ionising photons to the form

$$\frac{dF_{\text{n}}}{dr} \simeq - \left[\alpha_{\text{HI}}^{*}(T_0) \, n_e(r) \, n_p(r) \right]. \tag{7.67}$$

In effect we have put $4\pi r^2 \rightarrow 4\pi R_{\text{HII}}^2$ in equation 7.59, and then cancelled this term on both sides of the equation.

Substituting for dF_{n}/dr from equation 7.56, and for $n_e(r)$ and $n_p(r)$ from equation 7.53, we obtain, after some algebra

$$\frac{dx}{dr} = -n_0 \bar{\sigma}_{\text{HI}} \frac{[1 - x(r)]^2 x(r)}{[2 - x(r)]}. \tag{7.68}$$

We can integrate equation 7.68 by separating variables

$$\frac{[2 - x(r)] \, dx}{[1 - x(r)]^2 x(r)} = -n_0 \bar{\sigma}_{\text{HI}} \, dr, \tag{7.69}$$

and obtain the result

$$2 - \frac{1}{[1 - x(r)]} + \ln \left[\frac{x^2(r)}{[1 - x(r)]^2} \right] = n_0 \bar{\sigma}_{\text{HI}} [r - r_{1/2}], \tag{7.70}$$

where the constant of integration, $2 + n_0 \bar{\sigma}_{\text{HI}} r_{1/2}$, has been chosen so that $r_{1/2}$ is the radius where $x = 0.5$. In other words, the displacement $r - r_{1/2}$ is to be measured from the point where the degree of ionisation is 50%. The variation of x with r is illustrated schematically on Figure 7.9.

Typical values of the left-hand side of equation 7.70, for representative values of x, range from ~ 5 for $x = 10\%$ to ~ -12.5 for $x = 90\%$.

These must be equated to the right-hand side of equation 7.70, i.e. $n_0 \bar{\sigma}_{HI}[r - r_{1/2}]$. In addition, the number flux of ionising photons, $F_n(r)$, can be calculated, normalised to the value at $r_{1/2}$, namely $F_n(r_{1/2})$. From equation 7.55, this is given by

$$\frac{F_n(r)}{F_n(r_{1/2})} = \frac{2x^2(r)}{[1 - x(r)]}. \tag{7.71}$$

This fraction ranges in value, typically, from ~ 0.02 for $x = 10\%$ to ~ 16 for $x = 90\%$.

If we define the thickness of the transition region, ΔR_{IF}, to be the distance between where the degree of ionisation is 90% and where it is 10%, we have

$$\Delta R_{IF} \equiv r_{0.1} - r_{0.9} = [r_{0.1} - r_{1/2}] - [r_{0.9} - r_{1/2}]$$

$$= \frac{[(5.3) - (-12.5)]}{[n_0 \bar{\sigma}_{HI}]} = \frac{17.8}{[n_0 \bar{\sigma}_{HI}]}. \tag{7.72}$$

Substituting $\bar{\sigma}_{HI} \simeq 7 \times 10^{-22}$ m^2, this becomes

$$\Delta R_{IF} \simeq \frac{2.5 \times 10^{22} \text{ m}^{-2}}{n_0} \simeq 0.0008 \text{ pc} \left[\frac{n_0}{10^9 \text{ m}^{-3}}\right]^{-1}. \tag{7.73}$$

Combining this result with equation 7.48, we have

$$\frac{\Delta R_{IF}}{R_{HII}} \simeq 0.0005 \left[\frac{\dot{\mathcal{N}}_{HI}}{10^{50} \text{ s}^{-1}}\right]^{-1/3} \left[\frac{n_0}{10^9 \text{ m}^{-3}}\right]^{-1/3}. \tag{7.74}$$

Evidently, the assumption that the transition region is very thin is fully justified, and it is reasonable to describe it as an ionisation front.

7.9 Expansion of an HII region

We shall assume that, when a massive star forms, the output of hydrogen-ionising photons $\dot{\mathcal{N}}_{HI}$ builds up instantaneously and then stays at a constant value during the main-sequence life-time of the star. In other words, the star 'switches on' abruptly. The subsequent evolution can then be discussed in terms of three consecutive phases. We start by analysing the initial static phase, then the intermediate phase, in which the HII region undergoes dynamical expansion, and lastly the final phase, in which the expansion halts. We note that some of the most massive stars may burn out before the final phase is reached.

7.9.1 Initial static ionisation phase

In the first phase, the gas is static and the hydrogen-ionising photons from the central star ionise the gas out to a radius, R_0, defined by

$$R_0 \equiv R_{\mathrm{HII}}(t = 0), \tag{7.75}$$

which is given by the requirement of overall balance. The rate at which hydrogen-ionising photons are emitted by the central star must equal the total rate of recombination within the HII region (see equation 7.48), hence

$$R_0 \simeq 1.7\,\mathrm{pc} \left[\frac{\dot{N}_{\mathrm{HI}}}{10^{50}\,\mathrm{s}^{-1}} \right]^{1/3} \left[\frac{n_0}{10^9\,\mathrm{m}^{-3}} \right]^{-2/3}. \tag{7.76}$$

Here n_0 is the density of hydrogen nuclei (in all forms) in the surrounding undisturbed neutral gas. We have assumed that this is also the density of protons and electrons, i.e.

$$n_p \simeq n_e \simeq n_0, \tag{7.77}$$

in the initial HII region. This will be true provided that the gas does not have time to move significantly during this first phase, as the ionising flux from the newly switched-on central star first sweeps over it. We can justify this as follows.

The time-scale on which the initial equilibrium is established (i.e. the time-scale for the hydrogen ionising output from the central star to come into balance with the total rate of recombination in the HII region) is of order the recombination time-scale – i.e. the average time that a free proton or electron has to wait before it finds a mate with which to recombine, given by

$$t_{\mathrm{recomb}} = \left[\alpha_{\mathrm{HI}}^*(T) n_0 \right]^{-1} \simeq 170\,\mathrm{yrs} \left[\frac{n_0}{10^9\,\mathrm{m}^{-3}} \right]^{-1}. \tag{7.78}$$

The time-scale for expansion of an HII region driven by its over-pressure is the sound-crossing time, given by

$$t_{\mathrm{expand}} \simeq \frac{R_0}{a_{\mathrm{HII}}} \simeq 1.7 \times 10^5\,\mathrm{yrs} \left[\frac{\dot{N}_{\mathrm{HI}}}{10^{50}\,\mathrm{s}^{-1}} \right]^{1/3} \left[\frac{n_0}{10^9\,\mathrm{m}^{-3}} \right]^{-2/3}, \tag{7.79}$$

where a_{HII}, the isothermal sound speed in the HII region, is given by

$$a_{\mathrm{HII}} = \left[\frac{k T_{\mathrm{HII}}}{\bar{m}_{\mathrm{HII}}} \right]^{1/2} \simeq 12\,\mathrm{km\,s}^{-1}. \tag{7.80}$$

In this equation we have put $T_{\mathrm{HII}} = 10^4$ K, and $\bar{m}_{\mathrm{HII}} = 10^{-27}$ kg.

We require $t_{\mathrm{recomb}} \ll t_{\mathrm{expand}}$, which reduces to

$$n_0 \gg 1\,\mathrm{m}^{-3} \left[\frac{\dot{N}_{\mathrm{HI}}}{10^{50}\,\mathrm{s}^{-1}} \right]^{-1}. \tag{7.81}$$

This is clearly satisfied by the neutral gas in regions like giant molecular clouds where massive stars, and hence HII regions, form.

7.9.2 The dynamical expansion phase

In the second phase, the dynamical expansion phase, the HII region expands due to its over-pressure. Since it expands initially at a speed $\sim a_{\mathrm{HII}} \sim 12\,\mathrm{km\,s}^{-1}$, which is much greater than the sound speed, a_{HI}, in the neutral gas outside the HII region ($a_{\mathrm{HI}} \sim 0.3\,\mathrm{km\,s}^{-1}$), the ionisation front at the edge of the HII region is preceded by a shock front, which sweeps the neutral gas up into a dense shell.

To simplify the analysis, we shall make the following assumptions: We assume that global ionisation balance is maintained throughout, i.e.

$$\dot{N}_{\mathrm{HI}} = \frac{4\pi R^3(t)}{3} \alpha_{\mathrm{HI}}^*(T_0)\, n^2(t), \tag{7.82}$$

so

$$n(t) = \left[\frac{3\dot{N}_{\mathrm{HI}}}{4\pi R^3(t)\alpha_{\mathrm{HI}}^*(T_0)} \right]^{1/2} = n_0 \left[\frac{R(t)}{R_0} \right]^{-3/2}. \tag{7.83}$$

We assume that the dense layer between the ionisation front and the shock front is always very thin, because the shock compression is very large (see below). Hence both fronts have radial speed $\dot{R}(t) \equiv dR/dt$.

We assume that the pressure in the shocked gas is approximately equal to the ram pressure of the undisturbed neutral gas flowing into the shock front at speed $\dot{R}(t)$

$$P_{\mathrm{s}}(t) = \rho_0 \dot{R}^2(t) = \frac{n_0 m_p}{X_{\mathrm{H}}} \dot{R}^2(t). \tag{7.84}$$

We further assume that the pressure in the shocked gas is approximately equal to the pressure in the HII region

$$P_{\mathrm{s}}(t) = P_{\mathrm{HII}}(t) = \frac{n(t) m_p}{X_{\mathrm{H}}} a_{\mathrm{HII}}^2. \tag{7.85}$$

Finally, we assume that, although the pressures in the shocked layer and in the HII region are functions of time, they are uniform. This means that the evolution must be sufficiently slow that there is time for pressure waves to even out changes as they occur. This is usually the case, except

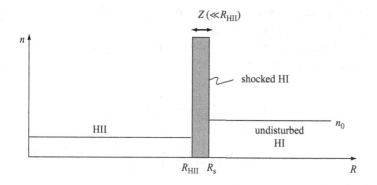

Fig. 7.10. The variation of density with radius through an HII region and the swept-up shell of neutral gas surrounding it.

in the very early stages of the evolution. See Figure 7.10 for a schematic representation of an HII region.

Eliminating P_s between equations 7.84 and 7.85, we obtain

$$n(t) = \frac{n_0 \dot{R}^2(t)}{a_{\mathrm{HII}}^2}. \tag{7.86}$$

Then eliminating $n(t)$ between equations 7.83 and 7.86, we have

$$\dot{R}(t) R^{3/4}(t) = a_{\mathrm{HII}} R_0^{3/4}. \tag{7.87}$$

This can be integrated to give

$$R^{7/4}(t) = R_0^{3/4} [R_0 + a_{\mathrm{HII}} t] \sim R_0^{3/4} a_{\mathrm{HII}} t. \tag{7.88}$$

The last expression in equation 7.88 gives the limiting behaviour at late times. This is an adequate approximation once

$$t \gg \frac{R_0}{a_{\mathrm{HII}}}$$

$$\simeq 0.17 \, \mathrm{Myr} \left[\frac{\dot{\mathcal{N}}_{\mathrm{HI}}}{10^{50} \, \mathrm{s}^{-1}} \right]^{1/3} \left[\frac{n_0}{10^3 \, \mathrm{cm}^{-3}} \right]^{-2/3}, \tag{7.89}$$

which is short compared with the main-sequence life-times of OB stars.

Using the limiting behaviour, we conclude that the time taken to expand by a factor $R_f / R_0 \sim 250$ (see equation 7.98 below) is

$$t_f \simeq \frac{R_0}{a_{\mathrm{HII}}} \left[\frac{R_f}{R_0} \right]^{7/4}$$

$$\sim 40 \, \mathrm{Myr} \left[\frac{\dot{\mathcal{N}}_{\mathrm{HI}}}{10^{50} \, \mathrm{s}^{-1}} \right]^{1/3} \left[\frac{n_0}{10^9 \, \mathrm{m}^{-3}} \right]^{-2/3}, \tag{7.90}$$

which is long compared with the main-sequence life-times of all but the lowest-mass OB stars. We conclude that the output of hydrogen-ionising photons from the central star is likely to run out whilst the HII region is still over-pressured and expanding.

Therefore, through most of its life-time the HII region's Strømgren radius is given approximately by

$$R(t) \simeq R_0^{3/7} [a_{\text{HII}} t]^{4/7} \simeq \left[\frac{3\dot{N}_{\text{HI}}}{4\pi \alpha_{\text{HI}}^*(T_0) n_0^2} \right]^{1/7} [a_{\text{HII}} t]^{4/7}$$

$$\simeq 5 \, \text{pc} \left[\frac{\dot{N}_{\text{HI}}}{10^{50} \, \text{s}^{-1}} \right]^{1/7} \left[\frac{n_0}{10^9 \, \text{m}^{-3}} \right]^{-2/7} \left[\frac{t}{\text{Myr}} \right]^{4/7} . \qquad (7.91)$$

The density in the HII region is then given by

$$n(t) = n_0 \left[\frac{R(t)}{R_0} \right]^{-3/2} \simeq n_0 \left[\frac{a_{\text{HII}} t}{R_0} \right]^{-6/7}$$

$$\sim 2 \times 10^8 \, \text{m}^{-3} \left[\frac{\dot{N}_{\text{HI}}}{10^{50} \, \text{s}^{-1}} \right]^{2/7} \left[\frac{n_0}{10^9 \, \text{m}^{-3}} \right]^{3/7} \left[\frac{t}{\text{Myr}} \right]^{-6/7} , \qquad (7.92)$$

and the mass of the HII region is

$$M_{\text{HI}}(t) = \frac{4\pi R^3(t)}{3} n(t) m$$

$$= \frac{\dot{N}_{\text{HI}} m}{\alpha_{\text{HI}}^*(T_0) n(t)} \simeq \frac{\dot{N}_{\text{HI}} m}{\alpha_{\text{HI}}^*(T_0) n_0} \left[\frac{a_{\text{HII}} t}{R_0} \right]^{6/7}$$

$$\sim 3000 \, M_\odot \left[\frac{\dot{N}_{\text{HI}}}{10^{50} \, \text{s}^{-1}} \right]^{5/7} \left[\frac{n_0}{10^9 \, \text{m}^{-3}} \right]^{-3/7} \left[\frac{t}{\text{Myr}} \right]^{6/7} . \qquad (7.93)$$

These relations hold for majority of the life-time of the HII region.

7.9.3 The asymptotic state

However, this can't go on forever. Eventually the expansion of the HII region will reduce its pressure to a value equal to the pressure in the undisturbed neutral gas, and the expansion will then halt (provided the star lives long enough). If at this stage the density and the radius of the HII region have values n_f and R_f, pressure balance with the undisturbed neutral gas requires

$$[2n_f] k T_{\text{HII}} = \left[\frac{n_0}{2} \right] k T_{\text{HI}},$$

$$\implies \frac{n_f}{n_0} = \frac{T_{\text{HI}}}{4 T_{\text{HII}}} \sim \frac{1}{4000}. \qquad (7.94)$$

Here we have put $T_{\text{HI}} \sim 10$ K, and $T_{\text{HII}} \sim 10^4$ K. The factors of 2 in equation 7.94 arise because in the ionised gas

$$n_{\text{total}} \sim n_p + n_e \sim 2n_0, \tag{7.95}$$

and in the neutral gas, which is mainly molecular hydrogen,

$$n_{\text{total}} \sim n_{\text{H}_2} \sim \frac{n_0}{2}. \tag{7.96}$$

In addition we must still have overall ionisation balance, giving

$$R_{\text{f}} = \left[\frac{3\dot{\mathcal{N}}_{\text{HI}}}{4\pi \alpha_{\text{HI}}^*(T)n_{\text{f}}^2} \right]^{1/3}. \tag{7.97}$$

Comparing equation 7.97 with equations 7.46 and 7.94, we have

$$\frac{R_{\text{f}}}{R_0} = \left[\frac{n_{\text{f}}}{n_0} \right]^{-2/3} = \left[\frac{4T_{\text{HII}}}{T_{\text{HI}}} \right]^{2/3} \sim 250. \tag{7.98}$$

However, this presupposes that the output of hydrogen-ionising photons from the central star is sustained for long enough for the HII region to expand by such a large factor, which is unlikely.

7.9.4 The swept-up neutral gas at the boundary of an HII region

As well as the ionised gas we should also consider the shell of neutral gas swept up between the ionisation front (IF) and the shock front (SF). With our assumption that the shell is very thin, the mass of the shell is

$$M_{\text{HI}}(t) = \frac{4\pi R^3(t)}{3} n_0 m - M_{\text{HII}}(t) \simeq \frac{\dot{\mathcal{N}}_{\text{HI}} m}{\alpha_{\text{HII}}(T_0)n_0} \left[\frac{a_{\text{HII}} t}{R_{\text{HII}}} \right]^{12/7}$$

$$\sim 15\,000\,\text{M}_\odot \left[\frac{\dot{\mathcal{N}}_{\text{HI}}}{10^{50}\,\text{s}^{-1}} \right]^{3/7} \left[\frac{n_0}{10^9\,\text{m}^{-3}} \right]^{1/7} \left[\frac{t}{\text{Myr}} \right]^{12/7}. \tag{7.99}$$

For a mature HII region, the mass of the swept-up shell of neutral gas at the edge is usually much greater than the mass of the ionised gas, and the final expression in equation 7.99 is obtained by neglecting the mass of the ionised gas.

As the surface-density of the shell grows it becomes increasingly gravitationally unstable, and eventually it should break up into collapsing fragments. These collapsing fragments may produce a new generation of stars (see Figure 1.11), and this is usually referred to as propagating, or triggered, star formation. If some of the new stars are massive, and therefore excite new HII regions, the process can repeat itself, and we speak of self-propagating star formation, such as is seen in Orion (see Figure 1.10).

Recommended further reading

We recommend the following books to the reader for further study.

Grebel, E. K. and Brandner, W. (2002). *Modes of Star Formation and the Origin of Field Populations*. Astronomical Society of the Pacific Conference Series, vol. 285. San Francisco: Astronomical Society of the Pacific.

Perez, E., Gonzalez-Delgado, R. M. and Tenorio-Tagle, G. (2003). *Star Formation Through Time*. Astronomical Society of the Pacific Conference Series, vol. 297. San Francisco: Astronomical Society of the Pacific.

Phillips, A. C. (2003). *The Physics of Stars*, 2nd edn. Chichester: Wiley.

Chapter 8
By-products and consequences of star formation

8.1 Introduction

In this chapter we discuss some of the phenomena observed as a consequence of star formation. We describe some of the phenomena surrounding star formation, such as discs, outflows, and binary and multiple stars, and we discuss the difference between hydrogen-burning stars and brown dwarf stars.

We then go on to detail some of the larger-scale consequences, such as how star formation affects the host galaxy in which it occurs. In this context we also discuss starburst galaxies and galaxy mergers. Finally, we outline current understanding on when the major epoch of star formation occurred in the Universe.

8.2 Circumstellar discs

In Chapter 6 we discussed accretion onto protostars. In particular, we discussed spherically symmetric accretion. However, if the material accreting onto a protostar has angular momentum (and in general it does), the infall is not spherically symmetric, nor is it direct. Instead, the material accumulates in a circumstellar disc, and then spirals inwards onto the equator of the star on a time-scale determined by the efficiency of the processes which redistribute or remove the angular momentum in the disc. Such a disc is often termed an accretion disc. We also mentioned this in Chapter 7 as a method for increasing the accretion onto a high-mass protostar in the context of significant radiation pressure potentially halting the accretion.

8.2.1 A model accretion disc

The processes which might redistribute angular momentum include magnetic and gravitational torques and turbulent viscosity, and are at best, poorly understood. However, we can still obtain a working model of an accretion disc, parameterised by the mass M_* of the central star and the accretion rate $\dot{M}_*$, which we assume to be constant.

We assume further that the gravitational field is dominated by the central star, so the disc is of sufficiently low mass for its self-gravity to be neglected. We also assume that the evolution is quasi-static, in the sense that the inward radial velocity component of the matter in the disc is much smaller than its orbital speed, and that the disc is supported centrifugally. Consequently we can put the orbital speed equal to the Keplerian value

$$v(r) \simeq \left[\frac{GM_*}{r} \right]^{1/2},$$ (8.1)

where $v(r)$ is the velocity at radius r, and M_* is the mass of the central star. This means that the kinetic energy per unit mass is equal to half the magnitude of the gravitational potential energy per unit mass

$$\frac{v^2(r)}{2} \simeq \frac{GM_*}{2r}.$$ (8.2)

It follows that the rate of release of gravitational potential energy between $r + dr$ and r (where dr is a small increment in r) is given by

$$\dot{M}_* \left[\frac{GM_*}{r} - \frac{GM_*}{(r + dr)} \right] = \frac{GM_* \dot{M}_* dr}{r^2},$$ (8.3)

where $\dot{M}_*$ is the rate of increase of mass of the star, i.e. the accretion rate onto the stellar surface. Half of this energy is spent increasing the orbital kinetic energy of the inward spiralling material (see equation 8.2). The other half has to be radiated away, or removed by some other mechanism.

8.2.2 Temperature profile

If we assume that the energy is removed by being radiated locally from the two sides of the disc, from the annulus between r and $r + dr$ (i.e. we neglect radial energy transport), and if we further assume that the disc is optically thick in the radiation it is emitting, then we can put

$$\frac{GM_* \dot{M}_* dr}{2r^2} = 4\pi r \, dr \, \sigma_{SB} T_s^4(r),$$ (8.4)

where $T_s(r)$ is the surface temperature of the disc at radius r, and σ_{SB} is the Stefan–Boltzmann constant. Therefore

$$T_s = \left[\frac{GM_*\dot{M}_*}{8\pi\sigma_{SB}r^3} \right]^{1/4}, \tag{8.5}$$

$$\implies T_s \propto r^{-3/4}. \tag{8.6}$$

Here $4\pi r\,dr$ is the surface area of the disc between radii r and $r + dr$ (both upper and lower sides of the disc contribute).

Given M_* and $\dot{M}_*$, we can calculate the emitted spectrum of the disc. The bolometric luminosity of the disc will be

$$L_{bol} \sim \frac{GM_*\dot{M}_*}{R_*}. \tag{8.7}$$

If the disc is not optically thick, it will need to be hotter to radiate the same amount of energy.

8.2.3 Flared discs

Some discs around protostars have spectra which are compatible with these theoretical predictions. Other discs have spectra which suggest that the temperature profiles in their discs are shallower, such as $T_s \propto r^{-1/2}$.

There are various ways of explaining this. For instance, if the disc is flared (its thickness increases with radius) then the disc may be heated by radiation from the central star and the surface temperature of the disc should fall off more slowly than $r^{-3/4}$. We return to the theme of disc evolution in Section 8.4 below, but first we discuss another by-product of star formation.

8.3 Bipolar outflows

Protostars often have bipolar outflows along their rotation axes – perpendicular to their circumstellar discs (see Figure 8.1). The mechanism driving these outflows has not yet been uniquely identified, but the outflows take a variety of forms.

On scales $\sim 10^{14}$ m (very near the star), there are jets. These are narrowly collimated beams of high-velocity material. On scales $\sim 10^{15}$ m, there are objects known as Herbig–Haro (HH) objects. These are small shock-heated knots where the jets impact the surrounding gas. On scales of $\sim 10^{16}$ m and larger, there are broad diffuse lobes of high-velocity CO emission. These are probably swept-up gas accelerated by the jets. Figure 8.2 shows some images of discs and jets around young stars.

Fig. 8.1. Image of a bipolar outflow.

Fig. 8.1. Image of a bipolar outflow.

Outflows were certainly not predicted before their discovery. In fact they came as something of a surprise when they were first observed, since everyone had expected to see infalling material in the formation of a star, rather than outflowing material. As we saw in Chapter 6, spectral signatures of infall have now been observed, but outflows have been known about for much longer.

The exact cause of outflows has not been fully explained in detail, but the broad picture is probably as follows. In Chapter 5 we discussed the angular momentum problem of star formation, whereby a cloud core spins up as it collapses and must shed most of its angular momentum if it is ever to form a star. The magnetic field is believed to be responsible for carrying away this excess angular momentum by being tied to the surrounding interstellar medium. This is known as magnetic braking.

The magnetic field lines can be pictured like a series of elastic bands tying the collapsing protostar to its surroundings. Figure 5.4 shows a schematic representation of this. As the protostar collapses and rotates the field lines are twisted into helical shapes. This has two effects: firstly by drawing the field lines closer together this increases the field strength; and secondly, the twisting of the field causes torsional Alfvén waves to travel along the field lines, rather like torsional waves on a string.

We discussed Alfvén waves in Chapter 4, and showed that they travel with a characteristic velocity of $B/(4\pi\rho)^{1/2}$. So the stronger the magnetic field, the higher the velocity of the waves.

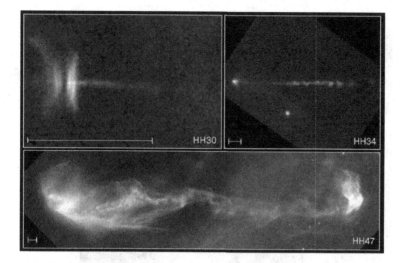

Fig. 8.2. Hubble Space Telescope images of young stars with oppositely directed jets, apparently emerging along the magnetic poles. In the upper left-hand image the disc around the young star is also visible.

It is possible that during protostellar collapse the magnetic field is twisted so tightly that high-velocity Alfvén waves are generated. Due to the strong coupling between the field and the matter – ambipolar diffusion is too slow to be a significant effect here (see Chapter 5) – some fraction of the infalling matter will be carried along with the Alfvén waves.

This matter will thus be carried in a clearly defined direction, along the magnetic poles of the protostar, at high velocity, due to the increased strength of the field. This is exactly what we appear to be seeing when we observe the bipolar jets and outflows from collapsing protostars.

Figure 8.2 shows examples of bipolar jets emerging from young stellar objects. In one image the disc can be seen edge-on, obscuring the central star. The jets emerge at right-angles to the disc, clearly along the axis of rotation of the system. Some debate remains as to whether the jets are launched from the very inner edge of the disc, or the stellar photosphere, and models exist which can account for either possibility. However, the broad picture of magnetic entrainment of matter in the jets is now generally agreed upon. In addition, the figure shows various bright points, or 'knots' along the jets, as well as Herbig–Haro objects, where the jets are interacting with the surrounding gas. None of these phenomena were expected before they were observed, showing that the study of star formation has proved to be a very unpredictable subject.

The most energetic example of an outflow that is currently known is in the star-forming region in the constellation of Orion. In this instance there is a large number of jets, apparently emerging from a central star. These objects are known popularly as the 'bullets' of Orion – see

Fig. 8.3. The 'bullets' of Orion. A violent star-formation event in which multiple jets, or bullets, appear to be emerging from a small region in the Orion star-forming complex.

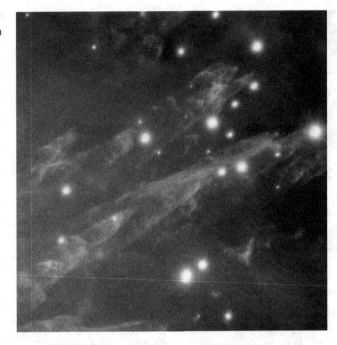

Figure 8.3. The star at the centre of such an explosive event must be of relatively high mass. The most likely candidate is known as the Becklin–Neugebauer (BN) object. This appears to be a very deeply embedded protostar, or cluster of protostars, in a very dense region. The theory of exactly how the bullets are formed, collimated and driven remains somewhat controversial, although clearly this is a very high-energy process.

The source of the energy of the outflows is a matter of debate. Many models have been formulated based on the energetic wind from the star. However, it is now becoming apparent that the only source of energy that is sufficiently large is the gravitational potential energy released during the accretion process. We showed in equation 8.2 that half of the gravitational potential energy released in an accretion disc is transformed into the kinetic energy of the material in the disc.

The amount of mass contained in a bipolar outflow is considerably less than that in an accretion disc. Therefore, even if the conversion efficiency factor between the kinetic energy of the disc and the kinetic energy of the outflow is only a few percent (depending upon the exact launching mechanism of the outflow), there is still sufficient potential energy released during the accretion process to generate the high velocities seen in bipolar outflows (up to a few hundred km s^{-1}). There is also growing evidence that bipolar outflow activity declines with declining accretion rate as the protostar evolves over time.

8.4 Disc fragmentation

It is now known that many stars have planetary systems orbiting around them. Our own Solar System is only one example. These planetary systems form from the discs of material around protostars that we discussed in Section 8.2 above. We have already discussed the likely temperature and density distributions in these circumstellar discs. We now consider their stability against fragmentation.

For this purpose we introduce a dimensionless parameter Q, which is known as the Toomre stability parameter. Q is defined by the equation (cf. equation 1.9)

$$Q = \frac{\sigma \kappa}{\pi G \Sigma}, \tag{8.8}$$

where σ is the root mean square velocity dispersion of the material in the disc and Σ is the mass surface density within the disc. κ is the epicyclic frequency, which is the frequency of oscillation of material to either side of its mean orbital radius. All of these quantities are functions of radius, and so Q is also a function of radius.

Note that in this analysis we are making the same assumptions as we made in Section 8.2 above, that the inward radial velocity of the material in the disc is negligible relative to its orbital velocity. For an infinitely thin disc the stability criterion requires $Q > 1$. That is, if $Q \lesssim 1$ there is a range of fragment sizes which are large enough for self-gravity to overcome the internal velocity dispersion, and small enough for self-gravity to overcome the internal spin. The smaller Q is, the larger this range of unstable fragment sizes becomes. For a finite thickness disc this changes slightly to $Q \lesssim 0.7$, but the result is the same.

Toomre's initial work was based on the discs of spiral galaxies. In fact, we discussed this criterion in Chapter 1 in respect of star formation in galactic discs. Nonetheless, his results hold true for circumstellar discs. The details of exactly how and when any given disc fragments must be modelled numerically on a powerful computer, and much research continues into the details of this process.

The ability of the disc to cool also has a bearing on whether a Toomre-unstable fragment will condense out. If t_c is the cooling time of the disc, and t_{orb} is the orbital period, at a given radius, then a fragment can condense out if

$$t_c \overset{<}{\sim} t_{orb}. \tag{8.9}$$

This is known as the Gammie criterion. If the cooling time is less than this critical value, the disc will fragment.

Once a fragment has formed within a disc, it is not inevitable that it will condense out. In particular, if it is unable to keep cool by radiating rapidly, then it will 'bounce' and re-expand. It is then likely to be sheared apart by differential rotation.

There are a number of processes that can destroy circumstellar discs. Photo-evaporation caused by heating from the central star is one such mechanism. We can consider a gravitational radius, r_g, such that

$$r_g = \frac{GM_*}{kT} \sim 100 \text{ AU}(T/1000 \text{ K})^{-1}(M_*/M_\odot), \quad (8.10)$$

where M_* is the mass of the central star and T is the temperature of the gas in the disc at any given radius. The significance of r_g is that at this radius the local sound speed is equal to the escape velocity from the system.

Hence for radii greater than this, the surface layers of the disc, which are warmed by the central star, can evaporate from the disc. In theory, gas in the disc at smaller radii remains gravitationally bound. More detailed analysis shows, in fact, that this changeover occurs at a radius closer to $0.2r_g$, but the principle remains the same.

Viscosity in the disc causes material to lose kinetic energy, and hence accrete onto the star itself. This is believed to be the dominant disc dispersal mechanism at radii very much less than r_g, whereas at very large radii the effects of other nearby stars must be taken into account. For example, in clustered regions such as Orion the external radiation incident on the outer parts of the discs causes them to evaporate on relatively short time-scales, effectively truncating the outer edge of the discs.

However, we do know that planets exist, so circumstellar discs must somehow form planets. We now look at some of the theories as to how this might occur.

8.5 Planet formation

The exact process that turns a disc of gas and dust into a system of planets such as our own Solar System is still very much a matter for debate. We summarise here some of the main ideas that have been proposed.

8.5.1 Formation of planetesimals

It is against a backdrop of disc destruction that planets must form. In particular, theories have difficulty explaining planet formation around massive stars, because their discs evaporate too rapidly for planet formation to occur. In discs where planets do form, it is necessary to consider different mechanisms in different regions of the disc.

In the inner disc, nearest to the star, no solid material can survive. Slightly further from the star, where the temperature is lower, some metals can condense out. Further still, silicate compounds can exist in solid form. These silicates are the dust grains that we have discussed previously (see Chapter 2).

One important boundary is known as the 'snow line'. This is defined as the radius at which volatile material such as CO and water can exist in solid form. At radii less than the snow line the volatiles remain in the gas phase, whereas at radii greater than the snow line the volatiles can form icy mantles on dust grains (see Chapter 4). Inside the snow line is where rocky planets form (e.g. Earth, Mars, etc.); beyond the snow line is where gas giants can form (e.g. Jupiter, Saturn, etc.).

At the densities typical of circumstellar discs, interstellar dust grains can grow to slightly greater sizes by a process of coagulation. This is where dust grains stick together, forming larger grains. The ratio of the photo-evaporation time-scale to the coagulation time-scale is an important consideration in planet formation.

The exact mechanism by which grains coagulate is uncertain. Firstly, the relative velocity between two grains must be very low, of order a few cm s^{-1}. For very small grains, which are well coupled to the gas, the Brownian motion of the gas may provide the impetus to cause low-velocity grain–grain collisions. For slightly larger grains, turbulent motions may be the cause, although too much turbulence can disrupt this process.

The grains may stick together by means of electrostatic forces. Alternatively, mild heating episodes of the grains may make their surfaces more 'sticky', causing them to coagulate. Beyond the snow line, where the grains have first acquired a mantle of molecular material in the form of water ice and CO ice, the dust grains may stick together like dirty snowballs.

The larger grains then settle towards the midplane of the disc under the effects of gravity due to their larger masses. This causes the dust-to-gas ratio to change in the midplane towards a greater dust fraction. This in turn increases the disc surface density in the midplane to the point where it exceeds that necessary for gravitational instabilities to develop. This occurs when Σ in equation 8.8 is sufficiently large for Q to be small enough to satisfy the Toomre instability criterion of $Q \lesssim 1$. We note that this process can only work if there is very little turbulence in the midplane of the disc.

This is the process by which it is believed that planetesimals form. Planetesimal literally means a piece of a planet. Typical sizes of planetesimals are kilometre-sized or larger. These are the building blocks of the rocky planets like the Earth, and also of the cores of gas giants such as Jupiter.

The role played by turbulence in this process is contentious. On the one hand, turbulent motions could disrupt the dust settling process, but on the other hand, turbulent eddies in the disc could act as vortices to gather material together to a sufficient degree for gravitational collapse to commence. This is known as trapping. This collapse could form planetesimals or even larger bodies. As yet there is insufficient observational data to judge.

Another problem in the formation of planetesimals is that of drag between the gas and the solids. At grain sizes of a millimetre or less the gas and dust orbit together. However, objects of centimetre to metre size experience significant drag from the surrounding gas. Models predict that such objects would rapidly spiral into the central star. Planetesimals of kilometre size or greater have sufficient inertia that this is not such a problem for them. However, the outcome is that whatever process takes material from millimetre sizes to kilometre sizes must occur rapidly.

8.5.2 Planetesimal growth

Once planetesimals have formed, then interactions between them are dominated by gravitational encounters. This introduces a random velocity component into the motions of planetesimals, in addition to their orbital velocities. Hence the likelihood of collisions is increased. The more massive planetesimals have the greater gravitational potential, and hence they grow the fastest.

This part of the evolution is known as the period of runaway growth. During this phase the largest planetesimals gradually accrete all of the smaller objects at similar orbital radii. This leaves what is often referred to as an 'oligarchy' of large planetesimals, each one dominant in its own radial regime in the disc. The subsequent phase of accretion of the remaining gas and dust is therefore known as the phase of oligarchic growth.

In this phase the planetesimals essentially clear out all of the material in their orbital path and create gaps in the disc corresponding to the orbital radius of each of the planetesimals. It is also possible in this phase that planetesimals can still interact gravitationally to merge and form more massive planetesimals. When the reservoir of dust in the disc has all been accreted onto the planetesimals and each is in a stable orbit, the result is a series of rocky planets such as we observe in the inner Solar System.

8.5.3 Giant planets

The giant planets have a slightly different formation mechanism. In fact there are currently three competing theories for the formation of massive planets such as Jupiter. The first theory has the rocky cores of giant planets forming in exactly the same way as the terrestrial planets.

However, because the giant planets form beyond the snow line, there is a significant amount of condensed volatiles in the neighbourhood of each planet to accrete onto it. So each planet acquires a large gaseous envelope.

The second theory predicts that a large-scale turbulent vortex forms at some point in the disc. This traps sufficient dusty material to allow planetesimal growth to proceed to the point of forming a rocky core. The turbulent vortex also traps volatile material and so a large gaseous atmosphere accretes onto the core.

The third idea invokes a large-scale gravitational instability in the circumstellar disc, which simply collapses to a giant planet. There are some theorists who say that this is not a mechanism for forming planets, but rather a mechanism for forming very low-mass stars – known as brown dwarf stars (see next section). Furthermore, in the current age where definitions of planets have become controversial, and the difference between a star and a planet has become blurred, it is perhaps useful to differentiate between stars and planets in terms of their formation mechanism. We shall therefore adopt the definition that a planet is a body which forms and grows by means of coagulation, while a star is a body which forms by gravitational instability. In this context we are clearly defining a brown dwarf as a star, and we believe that this definition allows for a clear distinction between stars and planets.

Once a planet has cleared out its orbit of all gas and dust it can no longer grow, other than by mergers. However, the residual disc can still interact gravitationally with the newly formed planet. The interaction is a tidal one in which angular momentum is transferred between the planet and the disc material. This can cause the planet to lose angular momentum and spiral into the star.

If the interaction is sufficient to cause the planet to lose some angular momentum, but not enough for it to spiral into the star, then it could cause a giant planet which was formed beyond the snow line to end up in an orbit much closer in to its parent star. This is known as migration. This mechanism has been invoked to explain some of the planetary systems that have been observed around other stars, where planets larger than Jupiter have been observed in Mercury-like orbits. These systems are known as hot Jupiters, and the theories of migration used to explain their presence are sometimes jokingly referred to as jumping Jupiter theories.

It must be noted that such systems may still be the exception, despite a number having been found. This is because many of the detection methods used to hunt for planets around other stars (extrasolar planets) have very strong selection effects in favour of finding more massive planets in close orbits. For example, the commonest method, involving radial velocity measurements, is most sensitive to the largest radial velocities, which are only produced by massive planets close to their parent stars.

8.6 Brown dwarf stars

In the regime of very low-mass stars we encounter objects known as brown dwarf stars, or brown dwarfs for short. A brown dwarf is a star whose mass lies below the limit required for the burning of hydrogen at its core. This limit occurs at a mass of roughly 0.075 $M_\odot$ (1.5×10^{29} kg), and is known as the hydrogen-burning limit. Below this limit a star is supported by electron degeneracy pressure, and does not reach a sufficient temperature at its core to burn hydrogen.

In Chapter 5 we derived the minimum mass of a star, based on thermodynamic considerations. We found this mass to be considerably below the hydrogen-burning limit for contemporary Population I stars. Therefore, there is no apparent theoretical restriction to the formation of brown dwarf stars, and in this section we discuss some of their observed properties and relate these to theories of brown dwarf star formation.

8.6.1 Brown dwarfs and planets

Given the low masses of brown dwarf stars, one might reasonably ask what the difference is between a brown dwarf star and a planet. This is a debate which has not been fully settled (although we offer one solution below). After all, the hydrogen-burning limit is only 80 times the mass of Jupiter (1.9×10^{27} kg), and planets up to $\sim$10 times the mass of Jupiter have been observed around other stars.

Originally the distinction was quite clear – a planet is a body which orbits a star, while a brown dwarf can exist in isolation. However, such a separation is no longer so clear. Some brown dwarfs have been found orbiting other stars (see below) and some planets have been found in isolation from any parent star – so-called free-floating planets. Hence the distinction is no longer clear-cut.

There is an inference that can be drawn from study of the stellar initial mass function (IMF), as discussed in Chapter 1. The IMF shows the relative numbers of high-mass and low-mass stars in any region. Observation of the IMF shows that it continues smoothly across the hydrogen-burning limit. There is no jump from one side of the limit to the other. Furthermore, brown dwarfs have broadly similar clustering properties and kinematics to hydrogen-burning stars. They can even have accretion discs and bipolar outflows like other stars (see above).

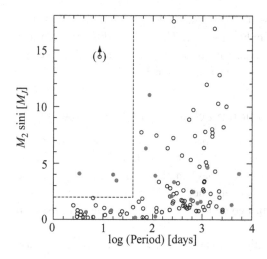

Fig. 8.4. A plot of mass vs orbital radius for low-mass stars orbiting stars of solar-type mass. The sparsely populated region of parameter space surrounded by the dashed box to the upper left of the figure is known as the brown dwarf desert.

Therefore, we refer to brown dwarfs as stars, and infer that brown dwarfs probably form in a similar manner to other stars, namely by gravitational instability. Planets, as we described in the previous section, form on a longer time-scale, by coagulation and accretion, and in general have a differentiated elemental composition. We note that this distinction is not universally accepted, but this is the differentiation that we make between a brown dwarf and a planet, in the context of star formation.

8.6.2 The brown dwarf desert

Now let us consider the binary properties of brown dwarfs. One early observation of brown dwarfs was related to their binary properties. It was noted that solar-type stars rarely have close brown dwarf companions. In fact for binary systems with separations ≤ 5 AU the frequency of companions in the mass range 0.01–0.1 $M_\odot$ is $\sim 0.5\%$.

Outside of this mass range, the frequency is much higher. Close binary companions of a few tenths of a solar mass are relatively common. Furthermore, planets around other solar-type stars (known as exoplanets) are also now known to be very common. The lack of close brown dwarf companions is known as the brown dwarf desert, and is shown schematically in Figure 8.4.

At wider separations brown dwarfs are observed orbiting solar-type stars. At separations greater than roughly 100 AU brown dwarf companions make up perhaps a few percent of the binary companions to solar-type stars. However, we note that this latter figure is based on smaller number statistics and may be less secure.

Binary stars in which both members of the binary system are brown dwarfs are more common. The fraction of binaries is at least as high as 10–20% and may be higher. Almost all have separations less than 20 AU, with a peak at only a few AU. This is very different from the binary statistics of solar-type stars discussed in Chapter 5, where we saw a very broad maximum peaking at more like 30 AU.

8.6.3 Possible formation mechanisms of brown dwarfs

A number of mechanisms have been proposed for the formation of brown dwarfs. In this section we discuss a few of the hypotheses. The most obvious idea is that they simply form in exactly the same way as solar-type stars. Namely, they result from the collapse of pre-stellar cores which in turn go on to form protostars, as we discussed in Chapters 5 and 6.

The problem with this idea is that the density of a very low-mass pre-stellar core has to be very high for it to be gravitationally bound. At typical densities in molecular clouds the Jeans mass M_J is around 1–3 $M_\odot$ (see Chapter 4). Recall that the Jeans mass scales with the inverse square root of the density ρ, such that

$$M_J \propto \rho^{-1/2}, \tag{8.11}$$

indicating that to obtain a Jeans mass two orders of magnitude lower, one needs to increase the mean density by roughly four orders of magnitude.

Moreover, all of the pre-stellar cores found so far have masses much greater than a typical brown dwarf mass, and so a way must be found to prevent the protostars that form in them from accreting more mass than the hydrogen-burning limit. We note in passing that there may be lower mass pre-stellar cores that have not been observed yet, simply due to the limited sensitivity of current observations, and that if such objects are discovered then this is no longer a problem. However, we proceed for now on the basis of what is currently known.

One solution would be if a pre-stellar core collapsed to form a triple system. In this case it is known that, through gravitational interaction, the lowest mass member of the triple is often ejected from the molecular cloud core, to leave a close binary of the remaining pair. If the ejection happened before the low-mass protostar had accreted 0.08 $M_\odot$, then it would not accrete any more mass and a brown dwarf would result.

Another solution is for a pre-stellar core to begin collapsing and to form a protostar in the normal way. But before the protostar can accrete beyond the hydrogen-burning limit the core is eroded by some external influence such as a nearby HII region (see Chapter 7). This would then prevent any further accretion by the protostar. If the external erosion occurred early enough in the protostellar accretion the result would be

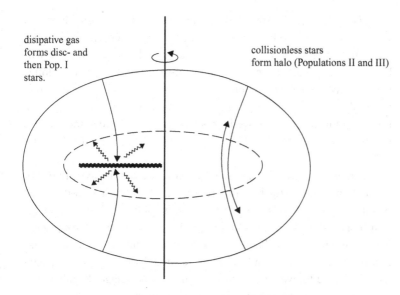

disipative gas
forms disc- and
then Pop. I
stars.

collisionless stars
form halo (Populations II and III)

Fig. 8.5. The collapse of a proto-galactic gas cloud.

a brown dwarf. However, we note that none of these suggestions is yet proven.

8.7 Galaxy formation

From the very small scale of brown dwarfs, we now turn our attention to the very large scale of an entire galaxy. The rate at which interstellar gas is converted into stars has a profound effect on the overall dynamics of a galaxy, and hence on its formation, structure and evolution.

Consider the formation of a galaxy. In the orthodox picture, a galaxy forms from a collapsing proto-galactic gas cloud.[†] As the proto-galactic gas cloud collapses, it flattens due to rotation. Eventually collapse orthogonal to the rotation axis is halted by centrifugal acceleration. However, the collapse continues parallel to the rotation axis (see Figure 8.5).

8.7.1 Stars

Any matter which by this stage has already been converted into stars is essentially collisionless. That is to say that the distances between stars are so large relative to their sizes that star–star collisions are extremely

---

[†] The conditions which must be fulfilled in the early Universe for these proto-galactic gas clouds to be created in the first place, and for their subsequent clustering properties to deliver what we see today, are amongst the most demanding constraints placed on cosmology by observation. Current theories invoke dark matter haloes to provide the gravitational potential wells necessary to seed the proto-galaxy. However, they are beyond the scope of this book.

rare. Therefore, as the stars fall into and through the midplane of the galaxy, they tend to be little influenced initially by other stars, and during the initial collapse they are all travelling inwards. However, there will be some exchange of kinetic energy between the stars, due to long-range gravitational interactions. This leads to the orbits of the stars being altered such that eventually at any time there are as many stars travelling inwards as are travelling outwards. Nevertheless, the stars do not dissipate any of their kinetic energy, and so they form a spheroidal halo.

Stars which form during the initial collapse of a galaxy are called Population II stars (for historical reasons). They are a well-observed population, and include most of the stars in globular clusters. Population II stars show signs of extreme age. For instance, the Hertzsprung–Russell diagram for a Population II globular cluster has a low-mass main-sequence turn-off, indicating that a significant fraction of the stars have had time to evolve past their main-sequence stage. Population II stars also tend to be on predominantly radial orbits.

Theoreticians have also proposed the existence of an even earlier population of stars, known as Population III stars. This population consists of stars formed before the formation of galaxies. No Population III stars have yet been observed, but if there is a Population III, these stars will also be collisionless, and so they could also end up in the spheroidal halo of a galaxy. The formation mechanism of Population III stars is very unclear. They are hypothesised to form from primordial hydrogen and helium, without any significant amount of heavier elements. Hence they are also known as low-metallicity stars.

Cooling a zero-metallicity primordial gas cloud sufficiently to allow it to become Jeans-unstable and collapse is a problem, since there are no dust grains or heavy elements. Cooling by hydrogen emission lines alone is very inefficient, and so a Population III pre-stellar cloud would be expected to heat up rapidly in the initial collapse phase. The reader will recall from Chapter 4 that the Jeans mass is strongly dependent on temperature – $M_J \propto T^{3/2}$ (equation 4.25). Hence, it would require a pre-stellar core with a great deal of mass to collapse to form a Population III star. This has led astronomers to believe that Population III stars could have masses up to $\sim 1000 \, M_\odot$. Such stars would go through their life-cycles very quickly and evolve to supernovae and ultimately black holes. This might explain why none have been observed.

8.7.2 Interstellar gas

Any matter which remains in the form of diffuse interstellar gas during the collapse of a galaxy converges on the midplane, where it runs into diffuse gas falling in from the opposite direction. The resulting collision causes a shock wave (see Chapter 4), which dissipates the bulk kinetic energy of the gas, converting it first into thermal energy, which is radiated away. The gas is then stuck in the midplane of the galaxy. It has no kinetic energy left to propel it up out of the bottom of the galaxy's gravitational potential well, so it relaxes to form a thin disc.

The stars which form subsequently in this disc are called Population I stars. They are characterised by being relatively youthful, having relatively high metallicity, and pursuing predominantly circular orbits. Population I star formation tends to be inefficient (in the sense that it only uses up the gas in the disc on a time-scale of order $3-10 \times 10^9$ years), and patchy (in the sense that at any time it is concentrated in a few locations – for instance, mainly in the spiral arms).

8.7.3 Ellipticals versus spirals

On this simple picture, elliptical galaxies were formed from proto-galactic gas clouds which converted virtually all their matter into stars before or during their initial collapse; consequently there was no diffuse gas left over to form a disc, and all the stars went into a spheroidal distribution. In contrast, spiral galaxies were formed from proto-galactic gas clouds which still had a significant component of diffuse gas left at the end of the initial collapse. This component formed a disc, whilst the component which was already in the form of stars formed a halo.

There are problems with this picture. For instance, it is now recognised that ellipticals are not normally flattened by rotation. They are flattened by an anisotropic velocity dispersion. This may also be true for the haloes of disc galaxies. Furthermore it is possible that mergers may play an important role in the formation of galaxies. In addition, galaxies can be stripped of their diffuse gas by the ram pressure of the intergalactic medium, as they move through the centre of a cluster of galaxies. And small galaxies can lose their interstellar gas just because they do not have a deep enough gravitational potential well to stop it being blown out by supernova explosions. We return to mergers in the next section, when we discuss starburst galaxies.

8.7.4 Spiral structure in disc galaxies

Once a disc galaxy has formed, star formation continues to play a major role in determining its external appearance. If one observes a disc galaxy in the optical, one sees typically a spiral structure, sometimes with a linear feature known as a bar in the centre. The bar and/or spiral arms are lit up by the light from the most massive newly formed stars. These massive stars only live a few million years, and so they are only seen close to the places where they form.

This is because the barred and spiral modes – which determine the redistribution of angular momentum in a disc, and hence the overall restructuring of the disc – are the principal triggers for star formation. The stars in the galactic disc produce a spiral modulation in the gravitational potential of a disc galaxy, and this in turn causes the interstellar gas to be compressed by a galactic-scale shock wave. This compression is presumed to be the trigger for the star formation seen along spiral arms, although it is still not understood to what extent the dissipation in the interstellar medium which accompanies star formation influences the development of these spiral modes, or their ability to transport angular momentum.

8.8 Starburst galaxies

In a normal spiral galaxy, such as our own Milky Way Galaxy, it has been estimated that stars are forming at a rate of only a few $M_\odot$ per year, averaged over the whole of the disc of the galaxy. But in regions of some galaxies this figure can be exceeded by orders of magnitude. Such galaxies are known as 'starburst galaxies'.

Typically a starburst galaxy displays this excess star-forming activity in and around the centre of the galaxy. The activity is seen not only as an excess in luminosity, but also in terms of its colours. A starburst galaxy will have a central region that is considerably bluer than a normal galaxy. In addition it will typically show excess emission in the UV and at higher energies such as X-rays. This emission is characteristic of very young, high-mass stars, and therefore of ongoing active star formation (since massive stars do not live very long and are therefore young). Figure 8.6 shows an optical image of the starburst galaxy M82.

Occasionally it happens that a starburst galaxy can be mistaken for an active galactic nucleus (AGN). An AGN is believed to be a massive black hole at the centre of a galaxy, which is accreting the surrounding material through a very energetic accretion disc (a higher-mass version of a circumstellar disc, such as was discussed in Section 8.2 above). But usually a starburst can be distinguished from an AGN by detailed spectral analysis of the central region of the galaxy.

A starburst is usually fuelled by a higher than normal rate of infall of gas towards the central region of the galaxy from the surrounding area. For some time it was unclear what drove these high infall rates, but now they are generally believed to be driven by the merger, or collision, of two galaxies.

Fig. 8.6. Optical image of the starburst galaxy M82.

When two galaxies collide, the stars within each galaxy are so far apart that they generally pass straight through without colliding. Kinetic energy is conserved and the interaction is elastic. However, the gas in the interstellar medium of each galaxy does collide, and when it does the collision is inelastic and kinetic energy is not conserved. This is because the collision causes shock waves in the gas, which heat up the gas and dissipate kinetic energy.

Figure 8.7 shows a picture of two colliding galaxies, illustrating how the material in one of them has been dragged out by the interaction with the other. Similarly, gas can be forced by a galaxy collision to fall into the centre of one of the galaxies at a much higher rate than would be caused by the gravitational field of one galaxy alone. A high rate of infall of gas into the centre of a galaxy triggers an episode of very active, massive star formation. This then manifests itself as a starburst episode within the life of the galaxy. Once the close encounter between the two galaxies is over and the infall rate returns to normal (or the reservoir of gas runs out) then the excessive star-forming activity also ceases and the galaxy returns to being a normal galaxy once more.

8.9 The epoch of star formation

In this section we look at how the star-formation rate in galaxies has varied throughout the history of the Universe. This is one aspect of star formation that has become a matter of much interest, as astronomers have asked the question of when in the history of the Universe the majority

Fig. 8.7. Optical image of two colliding galaxies, known as the Antennae. The effect of the collision can be to enhance the star-formation rate in the centre of one (or both) and create a starburst galaxy.

of star formation took place. Massive star formation can be traced by means of the high-energy radiation that it generates, in particular the UV radiation. Therefore, by tracing the UV emission as a function of the age of the Universe, we can trace the bulk of the star formation over time.

Due to the expansion of the Universe, distant galaxies are seen to be moving away at a velocity v that is proportional to their distance D

$$v = H_0 D. \tag{8.12}$$

This is known as Hubble's law and the constant of proportionality H_0 is known as Hubble's constant. The currently accepted value of H_0 is around 70 km s^{-1} Mpc^{-1} (although estimates of its value range from ~40 to 100 km s^{-1} Mpc^{-1}).

The recessional velocity leads to a Doppler shift of the emission from a galaxy to longer wavelengths. This is the famous red-shift of distant galaxies. The red-shift z is defined by the equation

$$\frac{\lambda_{\mathrm{obs}}}{\lambda_{\mathrm{em}}} = 1 + z, \tag{8.13}$$

where λ_{em} is the wavelength of emission when an object is at rest, and λ_{obs} is the observed wavelength. Hubble's law states that the further away the object, the faster it is receding. Doppler's law means that the faster an object is receding the greater its red-shift. Therefore astronomers often simply use red-shift z as an indicator of the distance of an object, since that is a directly measurable quantity. In addition, the further away an object is, the longer it has taken for its light to reach us, hence the further

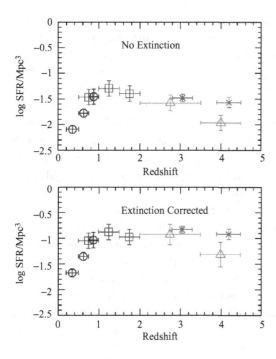

Fig. 8.8. A plot of the star-formation rate as a function of age of the Universe, as measured by the red-shift z. Zero is the present day and increasing z indicates decreasing age of the Universe.

back in time we are observing it. This means that z is also a measure of age.

So by measuring the UV emission from galaxies as a function of z, we can trace their star formation as a function of the age of the Universe. Due to the red-shift one must measure the UV emission in the rest frame of the galaxy; it may have been red-shifted into the optical wavelength regime for the observer, so the appropriate correction must be made.

It is therefore possible to estimate the total amount of star formation per unit volume of space by summing the total rest-frame UV emission in unit volume. One additional correction that must be made is that any given volume of space expands with the Universal expansion, and this must be taken into account. The corrected volume is known as the comoving volume to indicate that it is moving with the overall expansion.

Figure 8.8 shows the star-formation history of the Universe as a function of red-shift, sometimes known as the Lilly–Madau plot. Clearly there are many uncertainties involved in calculating the data-points on this plot. One of these is the fact that there is an unknown amount of dust extinction diminishing the UV radiation received from any one galaxy. An estimate can be made of this effect and measures taken to account for it. The far-infrared emission of a galaxy can also be used as a cross-check, since this is where the dust responsible for the UV extinction tends to re-emit most of the radiation.

Nevertheless, despite the uncertainties, Figure 8.8 appears to show that there was more star formation in the past than there is in the present-day Universe. There is a steep rise from $z = 0$ to $z = 1$. In addition there seems to be a broad peak of star-forming activity between red-shifts of roughly 1 and 2. This era is sometimes known as the epoch of star formation, as it was clearly an active time in the history of the Universe.[†]

The epoch of star formation may be caused by increased numbers of galaxy collisions at this time. Alternatively, it may be the epoch at which most elliptical galaxies formed most of their stars (there is very little star-formation activity in present-day elliptical galaxies). There are a number of hypotheses that have been put forward, although none have been proved. However, when we do understand the cause of the increased star-forming activity during the epoch of star formation we will understand considerably more about the evolution of the Universe.

There are so many aspects to the subject of star formation that one cannot cover them all in a short introductory text such as this. However, we hope that the study of this book will lead readers to follow up these other aspects for themselves, and in doing so, gain an insight into the manner in which star formation underlies so many other aspects of astrophysics.

Recommended further reading

We recommend the following textbooks for further reading on the topics covered in this chapter.

Binney, J. and Merrifield, M. (1998) *Galactic Astronomy*. Princeton: Princeton University Press.

Klahr, H. and Brandner, W. (2006). *Planet Formation: Theory, Observations, and Experiments*. Cambridge: Cambridge University Press.

Malbet, F. and Castets, A. (1997). *Low Mass Star Formation from Infall to Outflow*. International Astronomical Union Symposium, vol. 182. Dordrecht: Kluwer.

Reipurth, B. and Zinnecker, H. (2000). *Birth and Evolution of Binary Stars*. International Astronomical Union Symposium, vol. 200. Dordrecht: Kluwer.

Rowan-Robinson, M. (2004). *Cosmology*, 4th edn. Oxford: Oxford University Press.

[†] It should be noted that there is still some controversy over the higher-red-shift data-points in the upper plot of Figure 8.8, and some astronomers now believe that these points do not decline so steeply. Hence the peak may be even broader than is seen here. This is illustrated in the lower plot of Figure 8.8.

List of mathematical symbols

$a, \bar{a}$	radius of a dust grain and its mean
$\mathbf{a}$	constant relating temperature and energy density
a_0	isothermal sound speed
a_{grav}	acceleration due to gravity
a_{HII}	isothermal sound speed in HII region
a_{radn}	acceleration due to radiation pressure
A	area
A_{ji}	Einstein A-coefficient
A_V	visual extinction
α	infrared spectral index
α_H^*	recombination coefficient of atomic hydrogen into excited states
$\alpha_X(T)$	recombination coefficient at temperature T
$\mathbf{B}, B$	magnetic field strength
B_{ij}, B_{ji}	Einstein B-coefficients
$B_\nu(T)$	Planck function for a blackbody
β	dust emissivity index
$\beta_\nu(T)$	emissivity coefficient
c	speed of light in a vacuum
c_p	specific heat at constant pressure
c_v	specific heat at constant volume
C	a constant
C_{ij}, C_{ji}	collisional excitation and de-excitation coefficients
D	distance
$\mathcal{D}$	fractal dimension
ΔE	difference in energy
ΔS	mass converted into stars
$\Delta v, \Delta u_D, \Delta v_N$	velocity width or velocity dispersion
Δv_N	Natural line width
e, exp	base of normal logarithms, exponential
e^-	charge on electron
$\mathbf{E}$	electric field strength
E, E_i, E_j	energy, energy of levels i and j
$E(J)$	rotational energy of a molecule
E_m	magnetic energy

$\mathcal{E}M_{\rm H}$	emission measure of hydrogen
$\eta_{\rm SF}$	star-formation efficiency
f	compression factor
f_0, f_1	angular momentum of H in ground and excited states
$f_{\rm H}$	fraction of hydrogen in atomic form
f_λ	normalised monochromatic flux
$\mathbf{f_m}$	force per unit volume
$\mathbf{F}$	force
F	integrated flux
F_L	Lorentz force
$F_n(r)$	radial number flux of hydrogen ionising photons
F_ν	monochromatic flux density
$\mathbf{g}, g$	gravitational acceleration
$g(f)$	function of compression factor f
g_i, g_j	statistical weights of energy levels E_i, E_j
g_0, g_1	statistical weights of ground and first excited states
G	gravitational constant
$\mathcal{G}$	current total mass of interstellar material
$\mathcal{G}_0$	initial mass of interstellar material
γ	ratio of specific heats
γ_ν	photon of frequency ν
h	Planck's constant
$\hbar$	h divided by 2π
H_0	current value of Hubble's constant
$\mathcal{H}_{\rm comp}$	compressional heating rate
i	proton spin
I	integrated intensity
$\mathcal{I}$	moment of inertia
$I_{\rm CR}$	cosmic ray ionisation rate per unit volume
$\mathcal{I}_{\rm HI}$	rate of hydrogen ionisation per unit volume
I_ν	monochromatic intensity at frequency ν
J	molecular angular momentum quantum number
$\mathbf{J}$	electric current density
j_ν, j	monochromatic and integrated volume emissivity
k	Boltzmann constant
$\hat{\mathbf{k}}$	unit normal vector
K	a constant
$\mathcal{K}$	translational kinetic energy
κ	epicyclic frequency in a disc
$\kappa_1, \kappa_2, \kappa_3$	opacity normalisation constants
κ_d	dust mass opacity coefficient
κ_ν	monochromatic volume opacity
κ_V	volume opacity

l, L	length or path length through a medium
l	electronic angular momentum quantum number
l_ν	mean free path at frequency ν
$\mathcal{L}$	luminosity
L_{acc}	accretion luminosity
L_{BOL}	bolometric luminosity
L_ν	monochromatic luminosity
L_{SMM}	submillimetre luminosity
L_λ^*	monochromatic stellar luminosity at wavelength λ
$\lambda, \lambda_{\mathrm{em}}, \lambda_{\mathrm{obs}}$	wavelength, emitted, observed
λ_{max}	peak wavelength of blackbody function
m_d, m_{dust}	mass of a single dust grain
m_f	total angular momentum quantum number
m_i	proton spin quantum number
m_l	electronic magnetic quantum number
m_{p}	mass of proton
m_s	electron spin quantum number
m_X	mass of particle of species X
M	mass
M_c	critical mass
$M_{\mathrm{cl}}, M_{\mathrm{core}}$	mass of a cloud, clump or core
M_d	total mass of dust
M_{env}	envelope mass
M_{frag}	fragment mass
M_{J}	Jeans mass
$M_{\mathrm{max}}, M_{\mathrm{min}}$	maximum and minimum mass of a star
M_{vir}	virial mass
$\mathrm{M}_\odot$	solar mass
M_*	mass of a star
$\dot{M}_*$	accretion rate of a star
μ	mean molecular weight of gas
μ_{B}	Bohr magneton
n, n_{quant}	principal quantum number
n_{dust}	volume number density of dust grains
n_e	volume number density of electrons
$n(\mathrm{HI}), n_{\mathrm{HI}}$	volume number density of atomic hydrogen
$n(\mathrm{H_2}), n_{\mathrm{H_2}}$	volume number density of molecular hydrogen
n_i	volume number density of ions
n_p	volume number density of protons
n_X	volume number density of particles of species X
$n_{X;i}, n_{X;j}$	volume number density of particles in levels i or j
n_0, n_1	volume number density of particles in levels 0 or 1

n_0, n_f	initial and final volume number densities
n_{total}	total volume number density
$\mathcal{N}$	total number of molecules
N_c	number of clumps or cores
N_d	column density of dust grains
$\mathcal{N}_d$	number of dust grains
$\dot{\mathcal{N}}_H$	rate of emission of hydrogen-ionising photons
$N(\mathrm{HI})$, N_{HI}	column density of neutral hydrogen
$\mathcal{N}_M$	number of stars of mass M
N_X	column density of particles of species X
$N_{X;i}$, $N_{X;j}$	column density of particles of species X in level i, j
ν	frequency
ω	angular velocity
ω_0	initial angular velocity
Ω	solid angle
Ω_c	angular size of a cloud
Ω_G	gravitational potential energy
Ω_0	initial gravitational potential energy
P	pressure
$\mathcal{P}$	perimeter of a closed contour
P_m	magnetic pressure
$\phi(M)$	initial mass function of stars
$\phi(\nu - \nu_0)$	profile function
$\phi(R)$	potential function controlling radial excursions
$\phi_\nu(l)$	profile function at frequency ν along path length l
Q	Toomre stability parameter for a disc
Q_ν	dust grain emission efficiency
$\mathbf{r}, r$	position vector, radius (in disc)
r_{dust}	radius of a single dust grain
r_g	gravitational radius
r_J	Jeans length
r_n	cross-sectional radius of level of quantum number n
R, R_0, R_f	radius, initial and final
$\mathcal{R}$	Reynolds number
$\mathcal{R}_{\mathrm{HI}}$	rate of hydrogen recombination per unit volume
R_R	total recombination rate per unit volume
$\bar{\mathcal{R}}_{\mathrm{SF}}$	star-formation rate
$\mathcal{R}_X$	recombination rate of species X
ρ	density
ρ_d, ρ_{dust}	density of a single dust grain
ρ_0	initial density
s	electron spin

$\mathbf{s}$	radial direction vector
$\mathcal{S}$	surface
S_ν	source function
σ	standard deviation or velocity dispersion
σ_0	integrated cross-section of a transition
σ_{dust}	effective cross-section of a dust grain
σ_{HI}	cross-section of atomic hydrogen to an ionising photon
σ_{SB}	Stefan–Boltzmann constant
σ_X	cross-section of particle of species X
Σ	mass surface density (of a disc)
t	time
t_c	cooling time
t_{disp}	dispersion time
t_{ff}	free-fall time
t_{KH}	Kelvin–Helmholtz contraction time
t_{MB}	magnetic braking time
t_{orb}	orbital period at a given radius
T	gas-kinetic temperature
T_{dust}	dust temperature
T_{ex}	excitation temperature
T_R	rotational kinetic energy
T_S	surface temperature
T_0	initial rotational kinetic energy
T_*	surface temperature of a star
τ	optical depth
τ_ν	optical depth at frequency ν
τ_{AD}	ambipolar diffusion time-scale
θ	angular size
$\mathbf{u}$	velocity
u	integrated radiant energy density
u_ν	monochromatic radiant energy density
u_{rad}	radial velocity
u_T	turbulent velocity
u_x, u_y, u_z	velocity components in x, y, z directions
u_0, u_1	systematic and random velocity components
U_ν	monochromatic radiant energy density
v	velocity
v_A	Alfvén velocity
v_{escape}	escape velocity
$v(r)$	orbital velocity at radius r
v_x, v_y, v_z	velocity components in x, y, z directions
V	volume

W_λ	equivalent width of a spectral line
x	fractional ionisation
$x(r)$	degree of ionisation at radius r
X	atomic species
$X_f, X_{f'}$	atomic species X in free or unbound state
X_g, X_i, X_j, X_k	atomic species X in ground (g) or excited (i, j, k) states
X_H	hydrogen mass fraction
$\mathcal{X}$	ratio of H_2 column density to CO integrated intensity
z	red-shift
Z	metallicity
Z_{dust}	dust mass fraction
Z_X	partition function of species X

List of figure credits

Abbreviations

A&A	*Astronomy and Astrophysics.*
AAS	American Astronomical Society.
ALTAIR	Altitude conjugate adaptive optics for the infrared.
ApJ	*Astrophysical Journal.*
ARA&A	*Annual Reviews of Astronomy and Astrophysics*
AUI	Associated Universities Incorporated.
AURA	Association of Universities for Research in Astronomy.
ESA	European Space Agency.
ESO	European Southern Observatory.
FORS	Focal reducer and low dispersion spectrograph.
HST	Hubble Space Telescope.
IPAC	Infrared Processing and Analysis Center.
IRAS	Infrared Astronomical Satellite.
ISO	Infrared Space Observatory.
JPL	Jet Propulsion Laboratory.
MNRAS	*Monthly Notices of the Royal Astronomical Society.*
NASA	National Aeronautic and Space Administration.
NATO ASI	North Atlantic Treaty Organisation Advanced Science Institutes.
NICMOS	Near-infrared Camera and Multi-Object Spectrograph.
NIRI	Near-Infrared Imager.
NOAO	National Optical Astronomy Observatory.
NRAO	National Radio Astronomy Observatory.
NSF	National Science Foundation.
RAS	Royal Astronomical Society
STScI	Space Telescope Science Institute.
VLT	Very Large Telescope.
WFPC2	Wide-field and planetary camera 2.

Chapter 1

1.1   Figure courtesy of Sandra Arlinghaus, University of Michigan.
1.2   Image courtesy of VLT/ESO and the FORS Team.

1.3   From Figure 1 of Nutter, D., *et al.* (2006). *MNRAS*, **368**, 1833–1842. Reproduced by permission of the RAS.

1.5   Image courtesy of NASA/ESA, M. Robberto and the Hubble Space Telescope Orion Treasury Project Team.

1.6   From Figure 1(a) of Axon, D. and Ellis, R. (1976). *MNRAS*, **177**, 499–511. Reproduced by permission of the RAS.

1.7   Image courtesy of NASA/JPL-Caltech/Gordon, K., Willner, S. and Sharp, N.

1.8   Image courtesy of Ferraro, F. R., Shara, M. and The Hubble Heritage Team (AURA/STScI/NASA).

1.9   Image courtesy of Robert Gendler. http://www.robgendlerastropics.com

1.10   From Figure 7 of Blaauw, A. (1991). In *The Physics of Star Formation and Early Stellar Evolution*, ed. C. Lada and S. Kylafis, NATO ASI Series, **342**. Dordrecht: Kluwer. pp. 125–154. Reproduced with kind permission of Springer Science and Business Media.

Chapter 2

2.4(a)   Image courtesy of NOAO/AURA/NSF.

2.4(b)   From Figure 2 of Ward-Thompson, D., *et al.* (2006). *MNRAS*, **369**, 1201–1210. Reproduced by permission of the RAS.

Chapter 3

3.1   Image courtesy of NRAO/AUI and The HI Nearby Galaxy Survey (THINGS), F. Walter, PI.

3.6   Image courtesy of Five College Radio Astronomy Observatory and P. Goldsmith.

Chapter 4

4.2   Image courtesy of NASA/IPAC Infrared Science Archive and IRAS, a US/UK/NL mission.

4.3   From Figure 1 of Solomon, P., *et al.* (1987). *ApJ*, **319**, 730–741. Reproduced by permission of the AAS.

4.4   Adapted from the data of Caselli, P. and Myers, P. (1995). *ApJ*, **446**, 665–686. Reproduced by permission of the AAS.

4.5   From Figure 2 of Falgarone, E. and Phillips, T. (1990). *ApJ*, **359**, 344–354. Reproduced by permission of the AAS.

4.6   From Figure 7 of Lada, C., *et al.* (1994). *ApJ*, **429**, 694–709. Reproduced by permission of the AAS.

4.7   From Figure 1 of Crutcher, R., *et al.* (2004). *ApJ*, **60**, 279–285.
        Reproduced by permission of the AAS.
4.8   From Figure 4 of Redman, M., *et al.* (2002). *MNRAS*, **337**,
        L17–L21. Reproduced by permission of the RAS.

Chapter 5

5.1   From Figure 1 of Larson, R. (1969). *MNRAS*, **145**, 271–295.
        Reproduced by permission of the RAS.
5.3   From Harvard–Smithsonian Center for Astrophysics Press
        release. J. Girart, R. Rao, D. Marrone and Harvard-
        Smithsonian Center for Astrophysics Submillimetre Array.
5.5   From Figure 2 of Ward-Thompson, D., *et al.* (2002). *MNRAS*,
        329, 257–276. Reproduced by permission of the RAS.
5.6(a) From Figure 9 of Ward-Thompson, D., *et al.* (2005). *MNRAS*,
        360, 1506–1526. Reproduced by permission of the RAS.
5.6(b) From Figure 3 of Ciolek, G. and Mouschovias, T. (1994). *ApJ*,
        425, 142–160. Reproduced by permission of the AAS.
5.8   From Figure 1 of Ward-Thompson, D., *et al.* (2007). In
        Protostars and Planets V, ed. B. Reipurth, D. Jewitt and
        K. Keil. Tucson, AZ: University of Arizona Press, pp. 33–46.
5.10   From Figure 1 of Wilking, B., *et al.* (1989). *ApJ*, **340**,
        823–852. Reproduced by permission of the AAS.
5.11   From Figure 7 of Duquennoy, A. and Mayor, M. (1991). *A&A*,
        248, 485–524. Reproduced by permission of the ESO.

Chapter 6

6.5   From Figure 6 of Stahler, S. (1983). *ApJ*, **274**, 822–829.
        Reproduced by permission of the AAS.
6.6   From Figure 1 of Andre, P., *et al.* (1993). *ApJ*, **406**, 122–141.
        Reproduced by permission of the AAS.
6.7   From Figure 6 of Andre, P., *et al.* (1993). *ApJ*, **406**, 122–141.
        Reproduced by permission of the AAS.
6.8   From Figure 1 of Ward-Thompson, D., *et al.* (1996). *MNRAS*,
        281, L53–L56. Reproduced by permission of the RAS.
6.10   From Hogerheijde, M. (1988). PhD Thesis, Leiden University,
        Holland.
6.11   Image courtesy of HST/NICMOS, and D. Padgett, W. Brandner
        and K. Stapelfeldt.

Chapter 7

7.1   From Figure 1 of Parsons, H., *et al.* (2009). *MNRAS*, **399**, 1506–1522. Reproduced by permission of the RAS.

7.2   From Figure 1 of Longmore, S. N., *et al.* (2006). *MNRAS*, **369**, 1196–1200. Reproduced by permission of the RAS.

7.3   From Figure 1 of Avalos, M., *et al.* (2006). *ApJ*, **641**, 406–409. Reproduced by permission of the AAS.

7.4   Image courtesy of NASA and the Space Telescope Science Institute.

7.6   From Figure 8 of Zinnecker, H. and Yorke, H. W. (2007). *ARA&A*, **45**, 481–563. Reproduced by permission of the authors.

Chapter 8

8.1   Image courtesy of NASA and the Space Telescope Science Institute.

8.2   Image courtesy of HST/WFPC2/STScI, and C. Burrows, J. Hester and J. Morse.

8.3   Image courtesy of Gemini Observatory/AURA in conjunction with the NIRI and ALTAIR instruments.

8.4   Adapted from Figure 1 of Udry, S., *et al.* (2003). *A&A*, **407**, 369–376. Reproduced by permission of the ESO.

8.6   Image courtesy of NASA, ESA, and the Hubble Heritage Team (STScI).

8.7   Image courtesy of NASA, ESA, and the Hubble Heritage Team (STScI/AURA)-ESA/Hubble Collaboration.

8.8   From Figure 9 of Steidel, C. C., *et al.* (1999). *ApJ*, **519**, 1–17. Reproduced by permission of the AAS.

Index

Printed in the United States
By Bookmasters